Edutech Enabled Teaching

Technology in education will increase student's engagement and will also improve digital literacy of all the stakeholders of education world. Convergence of technologies like big data, AI, Blockchain will assist in analysis of data for generating meaningful policies for the betterment of education world. Real time connections can be built with technology and the internet. Undoubtedly, this book arrangement will without a doubt be a motivation for all of them.

The primary goal of this book is to address the issues faced by teachers in the adoption of digital tools into their teaching and their students learning. This book also addresses the issues confronting educators in the integration of digital technologies into their teaching and their students' learning. Such issues include a skepticism of the added value of technology to educational learning outcomes, the perception of the requirement to keep up with the fast pace of technological innovation, a lack of knowledge of affordable educational digital tools and a lack of understanding of pedagogical strategies to embrace digital technologies in their teaching. This book presents theoretical perspectives of learning and teaching today's digital students with technology and proposes a pragmatic and sustainable framework for teachers' professional learning to embed digital technologies into their repertoire of teaching strategies in a systematic, coherent, and comfortable manner so that technology integration becomes an almost effortless pedagogy in their day-to-day teaching.

Some of the objectives are given below:

- Shares valuable insights into the influence of technology on teaching and learning in higher education
- Provides deeper insights on higher education and sustainability interact
- Studies innovations from various perspectives
- Investigates how the educators and students apply the unique innovative and emotional dimensions in modern age of learning
- Provides a timely overview of changes in education reforms and policy research globally
- Evaluates the problematic relationship between globalization, the state, and education reforms.

This edited book presents theoretical perspectives of learning and teaching today's digital students with technology and proposes a pragmatic and sustainable framework for teachers' professional learning to embed digital technologies into their repertoire of teaching strategies in a systematic, coherent, and comfortable manner so that technology integration becomes an almost effortless pedagogy in their day-to-day teaching.

Edutech Enabled Teaching

Challenges and Opportunities

Edited by

Manpreet Singh Manna
Balamurugan Balusamy
Kiran Sood
Naveen Chilamkurti
Ignisha Rajathi George

CRC Press
Taylor & Francis Group
Boca Raton London New York

CRC Press is an imprint of the
Taylor & Francis Group, an **informa** business
A CHAPMAN & HALL BOOK

First edition published 2023
by CRC Press
6000 Broken Sound Parkway NW, Suite 300, Boca Raton, FL 33487-2742

and by CRC Press
4 Park Square, Milton Park, Abingdon, Oxon, OX14 4RN

CRC Press is an imprint of Taylor & Francis Group, LLC

Library of Congress Cataloging-in-Publication Data

Names: Manna, Manpreet Singh, editor.
Title: Edutech enabled teaching : challenges and opportunities / edited by Manpreet Singh Manna, B. Balusamy, Kiran Sood, Naveen Chilamkurti, Ignisha Rajathi George.
Description: Boca Raton, FL : Chapman & Hall/CRC Press, 2023. | Includes bibliographical references and index. |
Identifiers: LCCN 2022011777 (print) | LCCN 2022011778 (ebook) | ISBN 9781032185200 (hardback) | ISBN 9781032185217 (paperback) | ISBN 9781003254942 (ebook)
Subjects: LCSH: Education--Effect of technological innovations on--India. | Educational technology--India. | Internet in education--India. | Web-based learning--India. | Education and state--Inida.
Classification: LCC LB1028.3 .E4234 2023 (print) | LCC LB1028.3 (ebook) | DDC 371.330954--dc23/eng/20220510
LC record available at https://lccn.loc.gov/2022011777
LC ebook record available at https://lccn.loc.gov/2022011778

ISBN: 9781032185200 (hbk)
ISBN: 9781032185217 (pbk)
ISBN: 9781003254942 (ebk)

DOI: 10.1201/9781003254942

Typeset in Minion
by Deanta Global Publishing Services, Chennai, India

Contents

Preface

THE EVOLUTION OF TECHNOLOGY in the education sector dates back 2,500 years ago. In ancient times, people communicated orally only. For example, in Greece, education was restricted to mouth only. It was in the 15th century when writing method came into existence and followed by chalkboards and blackboards in 18th century. At the time of World War II, the US Army has brought the concept of the projector to train their forces which was later replaced by a higher quality projector with some superior inbuilt features like in-built software programs. Written communications became popular at the time of the Bible and gained more popularity in 15th century by Europeans when there was a trend among policymakers toward raising the literacy standards. The concept of video and radio technology was pioneered by the BBC in 1924. Satellite technology made its debut in the 1980's with the objective to broadcast educational programs to needy people. India launched its first satellite device in 1983 to disseminate education regionally. It was only in 1981 with the launch of the first movable personal computer that a revolution came in the journey of quality education. This was followed by the MacBook Pro laptop by Apple in 1987. Hence technology was picking up speed, and the future of education lies in the hands of technology. Nowadays, almost everyone is glued to the internet. The birth of social media platforms, such as Twitter and Facebook, has facilitated learning to some extent. The internet has developed many ways of interaction, like webcams and Skype. For instance, students can do research using Google. The primary goal of this book is to address the issues faced by teachers in the adoption of digital tools in their teaching and their students' learning. This book also addresses the issues confronting educators in the integration of digital technologies into their teaching and their students' learning. Such issues include a skepticism of the added value of technology to educational learning outcomes, the perception of the requirement to keep up with the fast pace of technological innovation, a lack of knowledge of affordable educational digital tools, and a lack of understanding of pedagogical strategies to embrace digital technologies in their teaching. This book presents theoretical perspectives of learning and teaching today's digital students with technology and proposes a pragmatic and sustainable framework for teachers' professional learning to embed digital technologies into their repertoire of teaching strategies in a systematic, coherent, and comfortable manner so that technology integration becomes an almost effortless pedagogy in their day-to-day teaching.Some of the objectives are sharing valuable insights into the influence of technology on teaching and learning in higher education, providing deeper insights on how higher education and sustainability interact, studying innovations from various perspectives, investigating how educators and students apply the unique innovative and emotional dimensions in the modern age of learning, providing a timely overview of changes in education reforms and policy research globally, and evaluating the problematic relationship between globalization, the state, and education reforms. We hope this book will help to promote reform in the field of Education.

About the Editors

Manpreet Singh Manna started his teaching profession as Lecturer from Sant Longowal Institute of Engineering & Technology (SLIET) in 1997. He then attained the position of Director, All India Council for Technical Education (AICTE) under Ministry of HRD, Government of India and New Delhi on deputation from 2014 to 2018 and worked as Director of the 'SWAYAM MOOCs' project and Prime Minister Special Scholarship Scheme (PMSSS). Dr Manna also taught in AIT, Bangkok, for the January 2012 session as Visiting Associate Professor by MHRD, Government of India. Dr. Manna has published about 140 research publications for leading national and international journals and has participated in conferences of repute. He is the editor and reviewer for many international refereed journals and has delivered expert talks on various topics on national television channels and at institutions in India. His renown has taken him to countries like the United States, Switzerland, Russia, Thailand, Singapore, and China to chair sessions for invited talks, present papers, and perform teaching assignments.

Balamurugan Balusamy is currently working as Associate Dean-Student Engagement with Shiv Nadar University, Del-NCR, India. Shiv Nadar University is recognized as Institute of Eminence by the Government of India. He had served up to the position of Associate Professor in his 14 years of experience with VIT University, Vellore. He has completed his Bachelors, Masters and PhD degrees from top premier institutions. His passion is teaching, and he adapts different design principles while delivering his lectures. He has written and published over 80 books on various technologies and visited 15 plus countries for his technical discourse. He has several top-notch conferences in his resume and has published over 150 journal articles, conference papers, and book chapters combined. He serves on the advisory committee for several technology startups and forums and does consultancy work for the industry on Industrial IoT. He has given over 175 talks in various events and symposiums. He can be reached at balamurugan.balusamy@snu.edu.in.

Prof. Simon Grima is the Deputy Dean of the Faculty of Economics, Management and Accountancy, Associate Professor and Head of the Department of Insurance and Risk Management which he set up in 2015, and started and coordinates the MA and MSc Insurance and Risk Management degrees together with the B.com in Insurance at the University of Malta. Simon is also a Professor at the University of Latvia, Faculty of Business, Management and Economics and a visiting Professor at UNICATT Milan. He served as the President of the Malta Association of Risk Management (MARM) and President of the Malta Association of Compliance Officers (MACO) between 2013 and 2015, and between 2016 and 2018, respectively. Moreover, he is the chairman of the Scientific Education Committee of the Public Risk Management Organization (PRIMO) and the Federation of European Risk Managers (FERMA). His research focus is on governance, regulations, and internal controls, and he has over 30 years of experience varied between financial services, academia, and public entities. He has acted as co-chair and is a member of the scientific program committee on some international conferences and is a chief editor, editor, and review editor of some journals and book series. He was awarded outstanding reviewer for the *Journal of Financial Regulation and Compliance* in the 2017 Emerald Literati Awards. Moreover, Simon acts as an Independent Director for financial services firms, sits on risk, compliance, procurement, investment, and audit committees and carries out duties as a Compliance Officer, Internal Auditor, and Risk Manager.

Kiran Sood is a Professor at Chitkara Business School, Chitkara University, Punjab, India and has received her undergraduate and post graduate degrees in commerce from Panjab University, respectively, in the years 2002 and 2004. She earned her Master of Philosophy degree in 2008 and Doctor of Philosophy in Commerce with a concentration on Product Portfolio Performance of General Insurance companies in India in 2017 from Panjabi University, Patiala. She has published articles in various journals and presented papers in various international conferences.

Naveen Chilamkurti is currently acting Head of Department, Computer Science and Computer Engineering, La Trobe University, Melbourne, VIC, Australia. He obtained his PhD from La Trobe University. Dr Chilamkurti is the Inaugural Editor-in-Chief for the International Journal of Wireless Networks and Broadband Technologies launched in July 2011, and he has published about 165 journal and conference papers. His current research areas include intelligent transport systems (ITS), wireless multimedia, wireless sensor networks, and so on. He currently serves on the editorial boards of several international journals and is a senior member of IEEE. He is also an associate editor for Wiley IJCS, SCN, Inderscience JETWI, and IJIPT.

Ignisha Rajathi George is a passionate student-centered educationalist with more than 14 years of experience in the Computer Science and Engineering discipline. She aspires her vision as a learner, facilitates her mission as a teacher, and accomplishes her responsibilities as a mentor to enunciate learning interest and critical thinking by pushing her learners beyond their comfort zone to taste success. Her expertise revolves around Medical Image Processing, Artificial Intelligence and Soft Computing. She is ardently interested in research showing progress in publishing high impact Thomson Reuters/Scopus Journal papers in Springer, MDPI, etc., national and international patents, book publications, Conferences (international and national) and her mentees have won awards in various 'hackathons' at international and national levels. Her research contribution has not left untouched the Books and Book Chapters, having been published in Springer and Elsevier publications. She received her Bachelor of Engineering and Master of Engineering as a rank holder in the discipline of Computer Science and Engineering under Anna University, Chennai. She has also completed her Doctorate in the Faculty of Information and Communication Engineering at Anna University, Chennai. She is currently working as Associate Professor, Department of Computer Science and Business Systems at Sri Krishna College of Engineering and Technology, Coimbatore, Tamil Nadu.

Contributors

Sugandha Agarwal
Research Scholar
Dr. A P J Abdul Kalam University
Lucknow, India

Anuradha Agnihotri
Dev Samaj College of Education
Chandigarh, India

Nima Anil
The Phd Scholar
Gandhigram Rural Institute (deemed to be a university)
Dindigul, India

Mahima Bhardwaj
Central University of Punjab
Bathinda, India

Surbhi Bhardwaj
The Technological Institute of Textile and Sciences
Bhiwani, India

Asawari Bhave Gudipudi
MIT Art, Design, and Technology University
School of Education and Research
Pune, India

Bhavana Behal
Banaras Hindu University
Varanasi, India

Priyanka Chhibber
Lovely Professional University
Phagwara, India

Aashish Dhiman
NICE School of Business Studies
Shobhit Institute of Engineering and Technology, (deemed to be a university)
Meerut, India

Simon Grima
Associate Professor Head
Department of Insurance
University of Malta
Malta, Europe

Deepa Gupta
GL Bajaj Institute of Management and Research, Greater Noida, India

Mukul Gupta
GL Bajaj Institute of Management and Research
Greater Noida, India

Parul Johri
Amity Institute of Biotechnology
Amity University
Lucknow, India

Prateek Kalia
Department of Corporate Economy
Faculty of Economics and Administration
Masaryk University
Brno, Czech Republic

Kuldip Kaur
Lovely Professional University
Phagwara, India

Ram Krishan
Mata Sundri University Girls College
Mansa, India

Divya Nair
MIT Art, Design and Technology University
School of Education and Research
Pune, India

Geetika Madaan
Chandigarh University
Mohali, India

S. Kavitha Maithily
Associate Professor
Gandhigram Rural Institute (deemed to be University)
Dindigul, India

Neeta Mhavan
MIT Art, Design and Technology University
School of Education and Research
Pune, India

Aastha Sinha
Shri Ramswaroop Memorial University
Barabanki, India

Amrinder Singh
Jain University
Bangalore, India

Jashandeep Singh
Chitkara Business School,
Chitkara University
Rajpura, India

Sachidanand Singh
Vignan's Foundation for Science Technology and Research
Guntur, India

Simran
Bhai Gurdas Institute of Management and Technology
Sangrur, India

Samridhi Tanwar
IG University
Meerpur, India

Neha Vashistha
NICE School of Business Studies
Shobhit Institute of Engineering and Technology (deemed to be a university)
Meerut, India

CHAPTER 1

Innovative Pedagogy Practices

Surbhi Bhardwaj and Samridhi Tanwar

CONTENTS

1.1 INTRODUCTION

Education is a key element in shaping the life of a person. The concept of education evolves from time to time, depending on skill sets and the lessons taught. In the 21st century, where everyone is facing a high level of competition in the world, simple knowledge focusing on theoretical aspects isn't enough to beat the global competition. In this competitive society, teaching approaches have shifted to instruction based on the students' learning capacities. The Pedagogy of Teaching is merely a focus. Teaching methods have an important part in shaping students' learning experiences and motivating them to seek out new and more information. Teaching methods have evolved from simple information sharing to interactive immersive experiences (Table 1.1).

DOI: 10.1201/9781003254942-1

TABLE 1.1 Generation Development

Twentieth Century (20th) Generation	New Generation
• Perception formed by reading	• Perception: Iconic
• Reading books	• Display for perception
• Independent	• Connected
• Linear approach	• Use of hyper media
• Elaboration/Discussion	• Notice/Warning
• Passive school	• School as a game/activity
• Awareness about environment	• Discovery of new things
• State of ambiguity	• Mutual cooperation
• Performing single task at a time	• Multitasking at a time
• Real world	• World of fantasies and imagination

1.1.1 Origin

Another word for education is 'pedagogy'. It is a professional and scientific method of teaching. The word pedagogy comes from the Greek words 'paidos' and 'agogos' Both words have their own meaning: 'paidos' means boy or child, and 'agogos' means leader, thus signifying the one who leads the child to school. In other words, pedagogy is a term that refers to both the teaching profession and the science of education, as well as a college course.

1.2 RESEARCH METHODOLOGY

1.2.1 Objectives of the Research

i. To study the detailed concept of 'Education System'

ii. To study the role of AI in pedagogy

iii. To study the pedagogy practices in India

1.2.2 Research Design

To study the above objectives, we used descriptive research design, so that we can easily study in depth with descriptions of whole content.

1.2.3 Data Collection Method

The data are based on secondary collection. We used several sources to make our study better and more meaningful. The main sources of collected data are:

Published articles, research papers, published journals, government reports, newspapers, online content, available facts and figures, magazines, and websites.

1.3 ANALYSIS AND INTERPRETATION

1.3.1 Education System

1.3.1.1 Education in Ancient India

The people of ancient India valued their well-organized, disciplined, and value-based Ancient Indian Education system. During that era, traditional and religious knowledge was taught to students. The writing pads were tree barks and palm leaves, and most of the teaching was conducted orally by sages and academics (Mili, 2019).

With the introduction of the Gurukul System of learning, education in India became more relevant. Religion, philosophy, warfare, medicine, astrology were the principal subjects of study and knowledge. This system was based on guru-dakshina.

1.3.1.2 Education in the First Millennium

Some universities, such as Nalanda University, Takshashila University, and Vikramshila University, began to spread knowledge among students at the turn of the 5th century. Suddenly, grammar, astronomy, philosophy, Arthashastra, logic, literature, medicine, law, Hinduism, Buddhism, mathematics, and logic were the subjects of utmost interest. During the 18th century, the availability of education spread widely to schools in most of the villages of India. Medieval times also focused on the establishment of madrasas, literary societies, and libraries.

1.3.1.3 Education in Modern India

During the British era, the new education system was launched with the learning of a new language, i.e., English. In the 20th century, Lord Macaulay proposed a new idea of education for India, which based on traditional Indian thought, knowledge, interests, and moral systems.

Post-independence, the education system in India was largely controlled by the central and state governments jointly up until 1976. Since the beginning of the new millennium, i.e., the 21st century, new education policies and plans have been implemented, such as the free education policy for children under 14 years and the aim to spend 6% of GDP on education, with a greater emphasis on primary education (Mili, 2019).

1.3.2 Pedagogy

In theory and in practice, pedagogy refers to the technique by which teachers teach. Pedagogy is shaped by a teacher's teaching views and is concerned with the interaction of culture and different learning styles. The study of teaching methods and how they affect students is referred to as pedagogy.

1.3.2.1 Innovative Pedagogy

The study of innovative teaching methods, as well as an examination of how those innovations in teaching methods and student interaction affect the teaching process, will be studied in this chapter. For instance, some of the following pedagogic styles improve the teaching and engagement of students:

Crossover teaching: While this type of education does not involve the use of technology, it is an enriching experience for both students and professors. Learning takes place in a different set-up, largely informal, such as after-school learning clubs or field visits to exhibitions and museums. Here, faculty can make connections between instructional content and students' experiences. This lesson is enhanced and deepened by the addition of subject-related questions. Photographic projects, group assignments/projects and field trip notes are subsequently used to add to the classroom discussions (Chiu, 2019).

Smart boards: Teaching through this pedagogy brings life to classroom instruction by allowing students to participate and comprehend more deeply. This is accomplished by incorporating interactive and visual elements into the course content. Faculty make use of innovative multimedia tools to elaborate the content more effectively to the students. Smart boards empower the teacher by making the teaching content more visual. It transforms the teaching experience into an interactive, engaging, and collaborative experience.

Flipping classrooms: A different approach of teaching is flipping the classroom. These days, it is very popular and a large percentage of teachers consider this a dynamic pedagogy. By putting the onus of learning on the students, they become active participants in the learning process. Teachers are relegated to the role of service providers whereas the students are responsible for obtaining concepts and knowledge. Faculty encourage students to generate skills/knowledge, develop inferences, and fill the information gaps by using a variety of technological tools.

Teaching through collaboration: Another creative teaching strategy is to encourage student collaboration on diverse projects. Collaboration is a fundamental life skill required for all kinds of jobs and businesses in today's globalized society. Allowing kids to learn, study, and work in groups can assist teachers in nurturing this talent in the classroom.

Using virtual reality to teach: The teaching approach of virtual reality aids the students in learning by allowing them to engage with a 3D environment. Instead of boring pupils with a history class, teachers may employ 3D technology to study old civilizations, journey to outer space for a science class, travel to different nations during geography session, etc. Virtual Reality technology facilitates students with enormous opportunities to learn in an immersive environment that will leave an indelible mark on their minds. It makes learning more enjoyable for kids and helps them retain information for longer periods of time all of which are important considerations when evaluating effective classroom teaching approaches.

Teaching through 3D printing technology: 3D printing can be used as a teaching tool by teachers who are looking for new ways to teach. This technology is quickly gaining traction around the world, particularly in higher education institutions, where faculty use 3D printers to construct prototypes, thereby simplifying complicated topics/subjects. Teachers can employ 3D printers in lower-level classes to teach knowledge that was traditionally taught through textbooks, allowing pupils to develop a greater comprehension of the topics – specifically in science, technology, engineering, and math (STEM) courses.

Using cloud computing to teach: Using technology in the classroom allows faculty to experiment with novel teaching methodologies. Teachers can use cloud computing to save critical classroom resources: for instance, notes, videos, lesson plans, audio lessons, and assignment details. With the touch of a mouse, pupils can access these resources according to their convenience from their homes or other places, as if they are in the classroom. This approach ensures that students who miss the session due to unavoidable reasons, like illness, are kept up to date at all times. It eliminates the need for students to tote about hefty textbooks and allows them to learn when, where, and at their own pace (Herodotou et al., 2019).

Technology and innovative teaching methods: Using technology in the classroom exposes students to a wide range of stimuli and fosters an environment favorable to activity-based teaching. It makes the subject of the classroom more interesting by learning a fun activity. Instructors have a nearly endless collection of tools at their disposal to fulfill the needs of their pupils thanks to technological advancements (Edsys, 2018).

1.3.2.1.1 Digital Education System The practice of incorporating innovative and cutting-edge digital tools and technologies into learning and teaching is known as digital education, e-Learning, or Technology Enhanced Learning (TEL). During the lockdown imposed by the governments, there was a rise in the adoption of digital education. The following sections discuss the positive as well as negative aspects of digital education during the pandemic.

1.3.2.1.2 Positive Aspects Digital education is dismantling many of the hurdles that impede kids in rural India from receiving a decent education in physically constrained classrooms (Jena, 2020):

Move toward 'edutainment': COVID-19 was key to the transition from traditional education to digital education delivery. 'Edutainment', which refers to a combined form of learning, has become popular in educational institutions. It pushed all parties involved in the educational system, including instructors and students, to become more technologically literate. It offers a new way to deliver classes and assessments of learning, opening up enormous possibilities for big changes in pedagogy and curriculum design. It facilitates learners to interact with large numbers of individuals at the same time (Chakraborty and Prasenjit, 2020).

Increase in usage of learning management systems (LMS): LMS have become very popular among educational institutions. This provided a chance for all organizations focused on the development and improvement of LMS, specifically for educational institutes, and the usage of electronic learning management systems (Misra, 2020).

Quality of learning material: Besides saving paper, COVID-19 created circumstances where students all over the world who were previously unable to submit their study materials in hard copy could now do so. Now, 'soft' (digital) versions of study materials are used by students (Vergnaud, 2020).

Enhancement of collaborative work: This is the era of collaborative learning and teaching, which has taken on different forms. Teachers from all around the world can collaborate to learn and share their knowledge (Misra, 2020).

Improvement in virtual meetings: Online/virtual meetings, teleconferences, e-conferences, and webinars were in high demand as a result of the epidemic.

Digital literacy improvements: This era introduced many individuals to digital technology and encouraged them to study and utilize it, resulting in an increase in the digital literacy.

Worldwide exposure: The in-person constraints caused by COVID-19 resulted in greater use and innovation of the internet for learning – a single platform for all students.

Educators and students now have a convenient and reliable method for communicating and learning with their peers all over the world (Chakraborty and Prasenjit, 2020).

Better time management: During the pandemic, students who are engaged in online education are better equipped to manage their time.

Promotes Open and Distance Learning (ODL): The majority of pupils are requesting ODL mode due to the pandemic crisis. During this pandemic, students are encouraged to engage in self-learning options that allow them to learn from a variety of materials and adapt their learning to meet their specific requirements.

'Direct to device' technology is accessible at any time and from any location, and it helps students gain access to high-quality education.

It allows individuals to save time by allowing them to progress at their own pace, as well as money by avoiding 'hidden costs' of education, such as commuting charges (gas, parking fees).

Working students will be able to avoid limiting their work schedules by not having to attend a specific class at a specific time, allowing them to earn more money.

Students can save more time and money with the flexibility of online courses, allowing them to learn with a purpose and instilling a feeling of self-confidence in them. As a tool, social media can be utilized to improve e-learning by making it more engaging, relevant, and culturally diverse. This also allows them the ability to ask questions and receive multiple responses in real time. Students can give and receive feedback on each other's assignments, as well as collaborate to generate conveniently accessible information. Students can use social media to be informed about current events, concerns, issues, social activities, and potential employment opportunities. In today's fast-paced world, the link between classroom-based learning and social media is critical.

Remote learning has certain requirements, such as recording lectures, uploading simple extracted files that are relevant to study, and so on.

1.3.2.1.3 Negative Aspects Educational activity hampered: Classes were halted and even exams were postponed at various levels, leading educational events and activities to be impeded. Annual examinations and entrance tests have already been postponed by various boards. A longer time period than expected was taken to complete the admissions process. Due to the lockdown, pupils lost around three months of the entire academic year of 2020–21, compromising educational continuity thus making it hard for pupils to continue schoolwork after a long hiatus.

Impact on employment: COVID-19 caused the majority of the recruitment to be postponed. Student placements suffered as a result of companies delaying student hiring. As a result of the epidemic, the percentage of unemployment is projected to climb. At present, there are no vacancies in government or the public sector in India, and young graduates who were offered jobs by private companies are concerned about their job stability. The jobless rate in India climbed from 8.4% (mid-March) to 23% (early April), according to the Centre for Monitoring Indian Economy. A hike in the unemployment rate (30.9%) in the urban sector was also recorded (Mahanand, 2020). When unemployment rises, people battle for food rather than schooling, which leads to a decline in education.

Unprepared students/teachers for online education: During the lockdown phase, a large percentage of students and teachers, unfamiliar with online classes, were not mentally prepared for the rapid transition from physical classes to online education. Due to the lack of a specialized online education platform, most faculty conducted classes through video platforms like Google Meet, Zoom, etc., which do not replicate actual online education.

The fall in employment opportunities at the global level: Several people lost their employment in foreign countries; on the other side, graduates may be unable to find work outside of India due to COVID-19 restrictions. Due to COVID-19, a large percentage of Indians returned, as they lost their jobs abroad (Chakraborty and Prasenjit, 2020). As a result, recent graduates who will be entering the labor market soon may have trouble finding acceptable jobs. Due to the lockdown, many students who had previously gotten employment through campus interviews were unable to start working. Indians who worked in other countries lost their employment. Because of the present pandemic crisis, fresh graduates in India are also concerned that job opportunities from corporate sectors would be withdrawn (Televisory, 2020).

Increased parental responsibility for their children's education: While the majority of educated parents are capable of teaching, guiding, and motivating their wards, some parents themselves lack the essential education to guide, teach, and encourage their wards at home.

Untrained teachers: The biggest issue that the Indian education system faced was a lack of well-trained teaching staff. Many teachers are unfamiliar with technological advancements. They needed high-quality, in-depth expert training to quickly improve their teaching style and pedagogy in accordance with e-learning materials.

Passive learning process: The student is responsible for digesting the information supplied on their own terms in passive learning. Students acquire information from the instructor and internalize it through passive learning. 'The learner receives no feedback from the instructor', according to this method.

Connectivity issues: This is a major problem in India. It's characterized as an outage at one of our upstream network providers that disrupts connectivity to the host server. The learning process cannot be visualized in reality without sufficient quality and speed of connectivity.

Distractions during study and learning: Online learning generates interruptions between study and learning since it lacks the framework of a regular classroom. Students' primary sources of distraction are social media, texting, television, and family, all of which can divert attention away from the task at hand and reduce productivity (Tagsold, 2013).

1.3.2.2 Pedagogy Practices

Pedagogical practices are those used by educators to help youngsters learn for the rest of their lives. Through an evidence-based approach, the learning opportunities and experiences enhance spontaneity, creativity, and the implementation of theory to practice. In the contemporary education system, modern and new social-digital practices are being adopted by educators. The major differences between the pedagogic practices are listed in Table 1.2.

TABLE 1.2 Differences between the Modern and New Social-Digital Practices

Social-Digital Participation	School Practices
Knowledge creators	Knowledge acquirers
High usage of the Internet	Limited textbook content
Use of digital media in a flexible way	Adoption of traditional media and e-mail
Information Communication Technology (ICT) tools for the mind	Pure mental capability
Working on a computer screen	Pen and paper
Create and share in groups	Individual achievements
Large networks	A closed classroom community
Multitasking	Linearity and sequence

1.4 FINDINGS AND SUGGESTIONS

1.4.1 Findings

The education sector in India is set to grow significantly in the coming years, according to the Digital India program's objective of transforming India into a digitally enabled society and knowledge economy. The socioeconomic impact of technology-driven reach and easy access will be felt by Indian students.

Government Intervention: Several e-learning platforms, apps and portals, such as Swayam, e-Pathshala, DIKSHA portal, STEM-based games and many more have been established by the government to help students continue their learning throughout the pandemic.

The use of digital learning has increased dramatically with one billion people using mobile phones and over 200 million mobile phones connected to the internet. A list of some of the issues follows:

Quality standard of education.

Inadequate teaching styles and knowledge of teachers.

High distraction of student during e-class.

Lack of tech-savvy teachers.

1.4.2 Suggestions

It is very beneficial to think digitalized, design digitalized, and being digitalized. But there are some suggestions necessary to follow in order for the successful adoption of digitalized transformation:

To assure student learning outcomes and educational quality, the Indian government must prepare post-pandemic education and research programs.

Need more funds for more transparency.

Improved network infrastructure.

Provide e-education training for both teachers and students.

1.5 CONCLUSION

Change is the end result of all true learning. Change is a 'must' for the growth, and students are the pillar of a strong base of knowledge and the overall economy. No doubt, COVID-19 has been a curse for the global economy, but a strong warrior always grasps change as an opportunity. So, it is bliss for developing and developed countries like India to take the advantage from this period with the help of the transformation of traditional education into digitalized education. Covid-19 has demonstrated how the education system in India takes advantage of inequities. As a result, renewed commitments to the synergy between the commercial and public education sectors are required. In this setting, it is necessary to make education a public good, and digital innovation can assist in this endeavor.

REFERENCES

Chakraborty, I. and Prasenjit, M. (2020). COVID-19 outbreak: Migration, effects on society, global environment and prevention, *Science of the Total Environment*, Vol. 728, pp. 138882.

Chiu, S. K. (2019). Innovative experiential learning experience: Pedagogical adopting Kolb's learning cycle at higher education in Hong Kong, *Cogent Education*, Vol. 6, No. 1, pp. 1–16.

Edsys (2018). 15 best pedagogical strategies for innovative learning. Available at: https://www.edsys.in/best-pedagogical-strategies/

Herodotou, C., Sharples, K., Gaved, M., Hulme, M., Rienties, A. K., B., Scanlon, E., Whitelock, D. (2019). Innovative pedagogies of the future: An evidence-based selection. Available at: https://www.frontiersin.org/articles/10.3389/feduc.2019.00113/full

Jena, P. V. (2020). Impact of pandemic COVID-19 on education in India, *International Journal of Current Research*, Vol. 12, No. 07, pp. 12582–12586.

Mahanand, S. (2020). The impact of COVID-19 on education and education sectors, know here. Available at: https://educationasia.in/article/the-impactof-covid-19-on-education-and-education-sectors-knowhere

Mili (2019). Pedagogical reform in Indian school education: Examining the child-centred approach, *Journal of Philosophy of Education*, Vol. 52, No. 3, pp. 1–15.

Misra, K (2020). Covid-19: 4 negative impacts and 4 opportunities created for education. Available at: https://www.indiatoday.in/educationtoday/featurephilia/story/covid-19-4-negativeimpactsand- 4-opportunities-created-for-education-1677206-2020-05-12

Tagsold, J. T. (2013). Why aren't they paying attention to me? Strategies for preventing distraction in a 1:1 learning environment, *The Journal of Research in Education*, Vol. 23, No. 2, pp. 126–145.

Televisory (2020). COVID-19 pandemic disrupts India's face-to-face medium of teaching. Available at: https://www.televisory.com/blogs/-/blogs/covid-19-pandemic-disrupts-india-s-face-to-face-medium-of-teaching

Vergnaud, S. (2020). What does COVID-19 stand for? Available at: https://www.facebook.com/100547337259893/posts/what-does-covid-19-stand-fordr-sophie-vergnaud-mbbsdr-sophie-is-the-clinical-exp/552079415440014/

CHAPTER 2

Education in the Fourth Industrial Revolution

Challenges and Opportunities

Geetika Madaan and Amrinder Singh

CONTENTS

2.1 INTRODUCTION

The Fourth Industrial Revolution (4IR) is the state of the art, in which developing technology and trends like the Internet and the artificial intelligence affect our lifestyle (Sima et al. 2020). The 4IR has many consequences for acquiring skills and educational attainment (ELayyan 2021). Redesign of education systems and the development of strategic creativity/innovation

DOI: 10.1201/9781003254942-2

methods are only a few examples. The present study will describe and conceive the 4IR to accomplish this objective (Liao et al. 2018). In the second phase, the properties of the 4IR are determined. The effects of the 4IR on education are also discussed in depth. Lastly, 4IR's educational challenges and possibilities are addressed in depth. Qualitative research techniques have been employed in this study, leading to unclear methodology. Disruptive research methods will be utilized to evaluate the literature and secondary material collected via desktop research, such as documentary analysis and conceptual analysis.

2.2 TIMELINE OF THE INDUSTRIAL REVOLUTION

To date, industrial revolution (IR) has been the greatest important development in human history during the last three centuries. In the history of human civilization, the phrase 'industrial revolution' may be defined as a significant period of growth, change, or transition, beginning with the creation of machines, telephones, and electricity (Schwab 2016). We are now at the beginning of a new industrial revolution, and it will change public policy, how the government delivers services to different regions, and the social and economic elements of society. Transition from a preindustrial to an industrial civilization, modern economic development is characterized by a rise in gross domestic product (GDP) per capita that occurs continuously and substantially over time and is referred to as an 'industrial revolution' (ELayyan 2021).

Three distinct industrial revolutions have occurred throughout history on the planet, one in each century. By the Middle Ages and until 1780, the First Industrial Revolution (1IR) was underway (Mohajan 2019). Several projects were undertaken to enhance communication skills and introduce stock exchanges, which resulted in the creation of banks and other financial institutions as well as private investments during the First Industrial Revolution (Nakicenovic et al. 2019). The First Industrial Revolution occurred mostly in the United Kingdom. The advancements brought on by this IR led other European nations to quickly adopt the revolution as their own. Many civilizations, from low-income to high-income, were profoundly altered by the First Industrial Revolution, enabling European nations to emerge from their respective 'dark ages'. European countries rose to power after the revolution was a resounding success. As a result of the First Industrial Revolution, human civilization progressed one step further (Gleason 2018).

Early nineteenth-century events are included in the First Industrial Revolution. During this period, human societies progressed from farming to the use of mechanical devices. Manufacturing techniques changed forever when the steam engine was developed in 1698 (Ivaldi, Scaratti, and Fregnan 2021). This allowed communities around the globe to develop creative, innovative methods to conduct business and address social problems, with Europe and North America leading the way in this regard. These transformations resulted in the creation of new opportunities, responsibilities, and skills. Many nations have been successful in developing, expanding, and diversifying their economies due to the First Industrial Revolution. The First Industrial Revolution (1IR) was primarily concerned with infrastructure development. The 1IR is well-known worldwide with its numerous achievements, including improving living standards, creating new skills, and expanding urbanization (Ivaldi, Scaratti, and Fregnan 2021).

The early nineteenth-century technical advancements continued throughout the Second Industrial Revolution (2IR). Steel, chemicals, energy, electricity, and many other sectors all saw major advances during 2IR's time. Numerous industries were able to function and grow their operations due to the development of electrical power. Mineral prospecting was made possible as a result of this technical advancement as well. The utilization of electrically driven machinery was a distinguishing characteristic of the 2IR (Chigwada and Nwaohiri 2021).

The Third Industrial Revolution (3IR) began midway through the twentieth century and continues now with no indications of slowing down. The 3IR was propelled by advances in production, distribution, and energy. During the 3IR, nuclear power was developed and electronics became much more widely used. The 3IR began in Europe, and as a result, Europe's technological progress started to fall behind the rest of the globe (Jules 2019).

Before the beginning of the twentieth century, industrial revolutions were characterized by advances in technology. For a variety of reasons, technology development, deployment, and application have accelerated in recent years: human beings' lives have grown more reliant on technological advancements; society has gained new skills and capabilities as a result of technological advancements, which also alters individual lives. As a result, the business is presently reorienting itself toward the Fourth Industrial Revolution. People can move and interact among digital domains using technology to assist them in managing their lives. In order to manage their lives, 4IR is a phenomenon that has enabled individuals to travel and communicate across digital domains (Butler 2018).

In addition, the 4IR is concerned with the present digital revolution. The Fourth Industrial Revolution has opened up new possibilities for society. It is based on the successes of numerous previous revolutions, including the Industrial Revolution. The twenty-first century will provide a variety of problems that will need the development of innovative solutions. The 4IR has several upcoming technologies. The new revolution includes fresh perspectives, new possibilities, new creations, and innovations (Matthews, McLinden, and Greenway 2021). The new revolution is all about tearing down barriers. 'The next revolution', is 'denoted by a much more all-encompassing and mobile Internet, smaller and more powerful, but less expensive, sensors, as well as artificial intelligence and machine learning' (Schwab 2016). The usage of mobile devices has increased dramatically in the twenty-first century.

4IR is comprised of DNA sequencing, nanotechnology, renewable resources, and quantum computing. As a result, the new revolution extends well beyond the use of intelligent technology, computers, and other such tools. As a result of this convergence of physical, digital, and biological technologies and their interconnections, the Fourth Industrial Revolution differs significantly from previous revolutions. These discoveries occurred simultaneously, ushered in a new era of growth, progress, and discovery for the global population. There is a revolution taking place right now in all regions, industries, and dimensions of existence (Spottl and Windelband 2021).

Four industrial revolutions across the world are pushing technical development, and the employment of many technological aids, governments, and the private sector is expanding rapidly. Thanks to technological advancements, a plethora of fresh thoughts and ideas have now been brought to reality. In the development of the new era, the Internet of Things, virtual worlds, massive data, and artificial intelligence (AI) are examples of such concepts. The advancement of people's lives and the availability to the general public of commercial businesses and public services have another similarity.

'Industry 4.0' is another name for the Fourth Industrial Revolution that is mostly used in the business sector, although it is also widely used in other fields. Industry 4.0 is a phrase that is used to refer to value chain technology organizations (Oke and Fernandes 2020). Industry 4.0 is concerned with the development of increasingly digital systems and the integration of networks via artificial intelligence. The Fourth Industrial Revolution (Industry 4.0) will see new systems replace current ways of completing jobs with work done by technologies.

An overview of the Fourth Industrial Revolution is shown in Figure 2.1.

Cyber-physical systems, industrial IoT components, Internet-enabled service providers all fall under the 4.0 umbrella. However, the 4IR has many other components, namely, 3D printing, artificial intelligence, blockchain, quantum computing (Figure 2.2). Goods and services, production, and consumption have changed as a result of these technological advancements.

Aspects of information and communications technology (ICT) are also included in the 4IR. IT is extensively utilized in a variety of settings including industry, government, and civil society. It is possible to accomplish activities using ICT when computer software and hardware are utilized together (Elayyan 2021). During the Fourth Industrial Revolution, (ICT) will be widely used across all sectors of society. These conditions have prevailed all across the world for many decades and they are still present now.

2.3 COMPONENTS OF TECHNOLOGY ADVANCEMENTS IN THE FOURTH INDUSTRIAL REVOLUTION

2.3.1 Big Data

'Big data' is a term that refers to huge, organized, and unstructured amounts of complex information that cannot be processed by traditional traction processing equipment or algorithms. Big data is defined as information assets with a large volume, a high speed, or a varied collection of characteristics requiring innovative and cost-effective information processing methods. Improved decision-making and process automation may achieved by

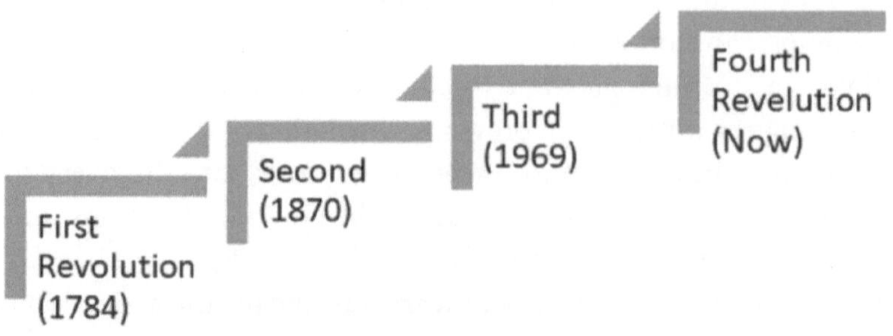

FIGURE 2.1 Timeline of the Fourth Industrial Revolution.

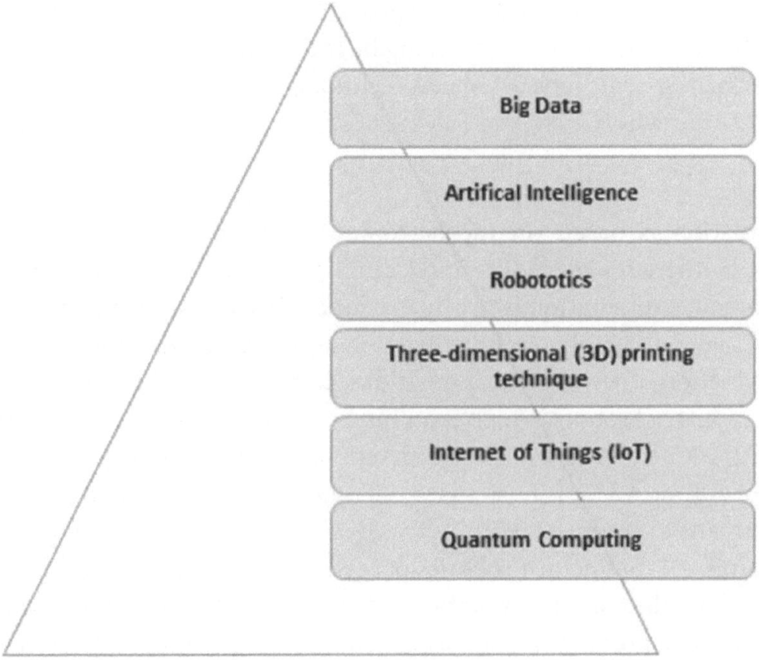

FIGURE 2.2 Components of technology advancements in the Fourth Industrial Revolution.

providing greater insight (Scepanovic 2019). Using big data platforms, companies will be able to analyze better and make decisions based on their available information. As the globe becomes increasingly connected, massive amounts of data are gathered from various devices (Sivarajah et al. 2017). These data must be examined in order to uncover any hidden information that may be there. The present ability to create and make information available is one of the main reasons behind the Big Data phase. Governments and companies depend on the volume of information produced using a range of technological methods when making decisions (Dash et al. 2019). Tracking the number of people who go to the local police station or provincial hospital, or the number of individuals who report a crime hotspot to the authorities are a couple uses for big data. Furthermore, huge quantities of collected data require the adoption of new technical innovations to store them all effectively (Chweya et al. 2020).

2.3.2 Artificial Intelligence

Artificial intelligence (AI) isn't a new concept; it's been debated for over 60 years. Complications relating to human intelligence may be performed by computers. However, AI has more cognition and capacity than humans (Collins 2021). In the field of artificial intelligence (AI) possesses greater intellect and capability. In artificial intelligence, the ability of computers to learn and react like humans is explained (Bjola 2021). As a result of new computer computers and instructions, computers are being trained to do tasks such as reasoning, thought processing, planning, learning, and sensing without human intervention. In other terms, AI is computer software that mimics human behavior. The usage of artificial intelligence in automobiles, Apple's Siri, and Google has grown worldwide (Siponen and Baskerville 2018). By analyzing huge amounts of data and patterns, computers learn to perform specific jobs using new technologies, and new technology is developed continuously (Zuiderwijk, Chen, and Salem 2021).

2.3.3 Robotics

Robotic automation is based on mechatronics, electrical engineering, and computer advancements (Sutherland 2020). The application of robotics varies in response to changes in the environment. Automation is firmly ingrained in sectors, such as automobile manufacture, which depends largely on the utilization of production equipment to achieve their goals. The production process in any industrial sector nowadays cannot be envisaged without the use of industrial robots, which is an important fact to remember (Gualtieri, Rauch, and Vidoni 2021). Machines are used in different sectors to increase production yields and a higher rate of output. Various businesses use robots to do jobs that are deemed too complicated for humans to complete daily.

For this reason, 4IR has helped robotics progress, since many countries desire the manufacturing process to be completely automated, or to allow 'smart automation'. Companies are trying to improve production, and as a result, they need more advanced robotics and artificial intelligence. As a result, new innovations and technologies will be developed to create robots that will outperform the previous generation (Rapanyane and Sethole 2020).

2.3.4 Internet of Things

The Internet of Things (IoT) is a novel idea that emerged as a result of technological advancements. Consider a world filled with a variety of electronic gadgets, such as refrigerators, coffee makers, television sets, mobile phones, and other similar items, all of which are connected to one another and to the Internet (Chweya et al. 2020). This is possible because of the Internet of Things idea, and various gadgets may interact and exchange information with one another thanks to this Internet connection. In the IoT, physical items are connected via embedded technology to communicate and sense their surroundings, as well as interact with the object's internal and external states. At the core of 4IR's design is the concept of four-dimensional interaction between machines and systems. Using IoT, people's daily lives and workplaces may be completely transformed (Liao et al. 2018).

2.3.5 3D Printing

The three-dimensional (3D) printing era has arrived. Since the advent of personal computers in the late 1970s, 3D printing has followed a development pattern similar to that of the technology. Additive manufacturing refers to the practice of making items from digital files by printing them in three dimensions (three-dimensional printing) (Chong et al. 2018). There are many uses for this cutting-edge technology in a wide range of industries such as medical and automotive production. Manufacturing aircraft and fabricating artificial organs from human cells are two examples of uses for 3D printing. This technology has been around since the 1980s, although there have been many improvements since then. While the technology has been around for a while, it has just lately taken off. Using 3D printing instead of traditional manufacturing is a more efficient and cost-effective solution (Teitel 2000).

2.3.6 Quantum Computing

It takes just a few days or hours, for quantum computing to tackle issues that would take thousands of years for conventional computers to solve (Rapanyane and Sethole 2020). New discoveries in health, energy, environmental systems, intelligent materials, and other fields are also made possible as a result of this technology. Quantum computers may help create scientific discoveries, lifesaving medicine, and early diagnostic machine learning, greater use of energy-efficient machinery and structural materials, as well as more comprehensive financial retirement plans and algorithms for quickly directing resources like ambulances (Rapanyane and Sethole 2020).

Modern civilization is characterized, in part, by the emphasis it places on the arts and sciences. The emphasis given to education is a defining characteristic of contemporary civilizations (Butt 2020). To reap the benefits of the 4IR in the digital age, countries must undergo a major transformation in their educational systems (Hoosain, Paul, and Ramakrishna 2020). Education, on the other hand, is prone to 'false problems'. Wicked issues require many explanations of important aspects and frequently include related difficulties. This article discusses a variety of important educational aspects. In essence, a new ICT strategy is needed to address some of the deadly issues presented by the 4IR.

2.4 IMPACT OF THE FOURTH INDUSTRIAL REVOLUTION REFORM ON EDUCATION

In order for the 3IR to create a 4IR educational plan, the 3IR will need to develop individualized teaching and a broad set of asynchronous instructional materials. A complex, dialectical, and exhilarating opportunity to make society a better place in the Fourth Industrial Revolution (4IR) is higher education. The 4IR has a variety of implications for a wide range of different life domains, and the result is that it provides both educational possibilities and difficulties (Teo et al. 2021). The education industry may be completely changed to provide answers to new problems by incorporating different components of the 4IR, namely, the Internet of Things, 3D printing, quantum computing, and artificial intelligence (Rotatori, Lee, and Sleeva 2021).

The impacts of 4IR in educational affairs have been on syllabus, education, and knowledge – rather than having robotic tutors – in the classroom. To put it another way, teaching and learning should be cross-sectoral. Educators and students from a variety of disciplines must get familiar with the many variables that influence the success of a 4IR application. Students pursuing degrees in fundamental and applied sciences must be aware of the political and social nature of the society in which they will be working. And it is essential for students studying both the natural sciences and the social sciences to be familiar with the fundamental concepts and operation of artificial intelligence (Tsiligiris and Bowyer 2021). Accordingly, the 4IR leads to the concept of a multidisciplinary area in which the sciences and humanities work together combining technology and science, 4IR, as well as the advancement of biotechnology and artificial intelligence, to address the significant challenges to human beings and their connection with the natural world. It is necessary to create 4IR liberal arts programming that takes into consideration socioeconomic dislocations in the 4IR. In general, the 4IR curriculum should address the political and social conflicts that have arisen as a result of the fast advancement of technology.

For example, new standards for digital pedagogy are needed to provide a theoretical foundation for online education and the increasing use of IA in education and learning. To be considered digitally literate, for participation in a global digital society, the acquisition of flexible skills is required. Utilizing the digital economy and generating new possibilities for employment, innovation, and creative expression are essential prerequisites for social inclusion (Howell, Perez, and Abraham 2021).

Considering the effects of change on education should be part of any digital education strategy. What a terrible situation: if students are not well prepared and inadequate resources are allocated, the quality of graduates may suffer. Education is especially vulnerable to negative consequences, particularly when it comes to quality measurements. The development and implementation of quality measurements, performance indicators, and educational outcomes in the context of educational reform and strategy are becoming more difficult.

Many people attribute R&D activities' drive for new technological advances to such a reality. Technologically based research has yielded many positive results. There are various kinds of technology-driven research and development, such as the use of mobile data collection capabilities and sophisticated big data analytics to discover statistical trends that are not immediately apparent; information searches, knowledge collection, organization, and discovery using artificial intelligence technology are all examples of this (Kocdar, Bozkurt, and Goru Dogan 2021).

Furthermore, sufficient abilities are required for the 4IR to be successful in the educational setting. Technology implementation, management, and working with modern technologies necessitates the use of collaborative skills. It is necessary to acquire the appropriate skills in order to accomplish the goal of getting the most out of new technologies. 35% of the abilities currently deemed essential in the present employment landscape will change shortly. Consequently, new skill sets and the use of new technologies will be needed for the new revolution. In order to succeed in industries like sales and manufacturing, employees must have some level of technical literacy. Tollbooth workers, for example, have been displaced while new professions (like social media experts) have been created by the advent of new technology.

Aside from that, 4IR has the potential to impact moral and ethical choices to be made, and they must be considered. Technology has various effects on people's everyday lives. Several dynamic changes are taking place due to the extensive use of contemporary technology in business, government, and other areas of life. In contrast to previous technological breakthroughs, such as the AI revolution, new technological advancements have the potential to have a more dramatic effects on society (Nagy et al. 2018); they can mobilize people on a scale that has never been previously witnessed. A large portion of the population may lose their jobs as a consequence of this. Many researchers, academicians, and policy makers, on the other hand, believe that the preceding assertion is incorrect. On the one hand, some people believe that the new revolution will expand the number of employment opportunities available to competent and prepared individuals. On the other hand, the argument is that widespread use of technology would eliminate existing *and* future job opportunities.

Alternative arguments assert that technology advances will lead to the elimination of existing and future jobs. Given the widespread use of new technology, such as robots, it is still unclear whether the decline in employment levels in the United States and other parts of the world may be attributed to either government or industry. As a consequence of the use of new technology, it has been suggested that human resource costs may rise, which is a concern that has been raised. In the field of education, ethics plays a crucial part in the success of students. Due to the increasing availability and accessibility of educational products, it is critical to emphasize the ethical limits that must be maintained in order to ensure the development of ethical values in education.

The precise ramifications of the 4IR on society are yet unclear (Lee et al. 2018). Unavoidable is the fact that education will be required to react swiftly to the advancement of technological innovation.

2.5 CHALLENGES OF THE FOURTH INDUSTRIAL REVOLUTION IN THE EDUCATION SYSTEM

There are challenges connected with 4IR that are distinct from those associated with new technology development and implementation (Figure 2.3). Careful planning is needed in order to minimize these risks. It is also essential to put in place new risk management systems and processes to protect the organization. As a result, new technologies have the potential to make a beneficial impact in the lives of those who use them (Brocal et al. 2019). In these new technical advances, there is a danger to humanity, and their negative repercussions must not be ignored.

When it comes to education, inequality is the first danger that 4IR poses. Unemployment and income distribution are societal problems in India. Distinct cultures are more prone than others to have high levels of crime, sexual violence, and unemployment related societal problems. Excessive inequality exists in the Indian education system, and new technology advancements have the potential to aggravate the situation (Sutherland 2020). There is a risk that only the richest parts of society will be able to afford new educational technology, leaving the poorest portions of society behind. This has been shown by the implementation of the previous three industrial revolutions, to mention a few examples. Many people today still do not have access to clean drinking water, electricity, transportation, or the Internet. As a consequence, growing inequality between the 'haves' and the 'have-nots' will lead to greater alienation, a lack of trust, and social instability.

It is necessary to take into account the human situation and social fairness. It is necessary to consider how technical advancements and shifting economic power have an effect on society at different socioeconomic levels. It is important to recognize the dangers that face us in an increasingly linked world and encourage multicultural understanding, respect for freedom, and human rights. As a result, it is necessary to acquire intercultural and interpersonal abilities.

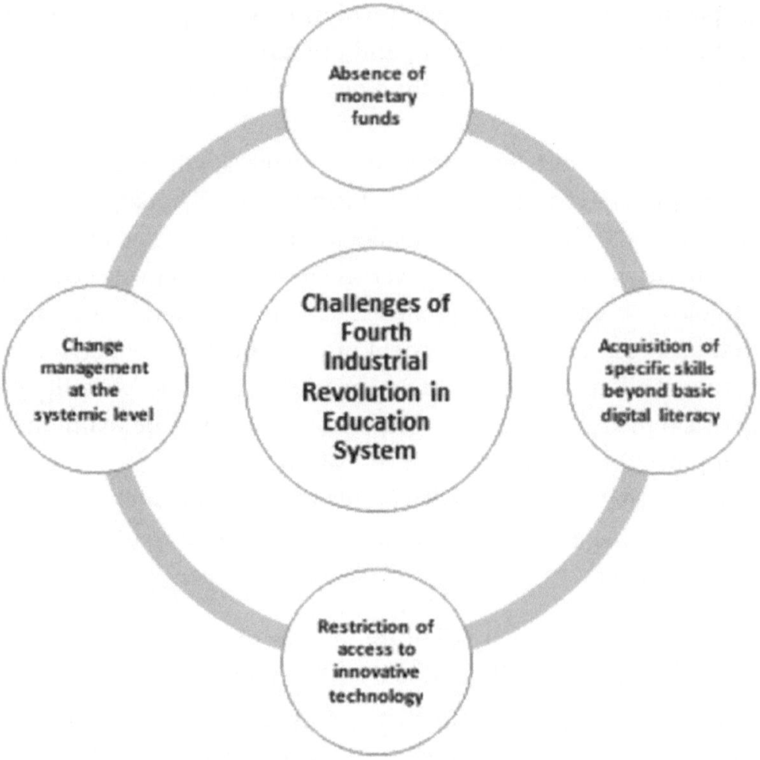

FIGURE 2.3 Challenges of the Fourth Industrial Revolution in the education system.

The National Development Plan, the Department of Basic Education's Action Plan, and the Department of Basic Education's Strategic Plan identify several challenges linked to education and technology development. The following challenges must be overcome:

- Identifying which instructors need pedagogical assistance;
- Conducting objective and timely evaluations of educational results;
- Imparting new knowledge and skills to teachers and students for the twenty-first century;
- Facilitating self-learning and discovery by providing learners with access to the Internet;
- Improving the quality of learning for students;
- Facilitating self-improvement and exploration;
- Accepting criticism and exposure to new concepts;
- Reducing administrative responsibilities, such as test-marking and absence reporting.

2.6 MAJOR CHALLENGES OF THE FOURTH INDUSTRIAL REVOLUTION IN THE EDUCATION SYSTEM

- **Lack of funding**

The absence of sufficient funding is one of the most significant obstacles to implementing 4IR in Indian education (Mohajan 2019). Education funding has risen in recent years, yet it is still inadequate to ensure that educational institutions can operate at full capacity. For example, this lack has increased university tuition and decreased research funding. The issue for educational institutions is to increase their investments in new technology advancements while also prioritizing the utilization of available resources. A significant amount of financial assistance is required in order for new technologies to flourish in educational institutions. Qualified instructors and technical infrastructure are the two most expensive components of training (Gleason 2018).

- **Marginalization of the Fourth Industrial Revolution by certain socioeconomic segments**

Another obstacle that the educational sector must overcome, in order to have an influence on the efficiency with which the 4IR plan is put into action, is the marginalization of the 4IR by certain socioeconomic segments. Even though the gross rate of socially marginalized student enrollment has increased since 1994, their actual number remains significantly lower than that of the general student population. Social justice is seen as being a duty of the university sector and for establishing a fair playing field in order to counteract the negative impacts of apartheid on socially marginalized students.

- **Lack of specific skills beyond basic digital literacy**

For the foreseeable future, personal and societal success in the 4IR will demand new skills. 4IR technologies and their uses need the acquisition of specific skills beyond basic digital literacy. In terms of these abilities, universities are at the forefront of the industry (Sivarajah et al. 2017). Being barred from attending university will prevent from acquiring these skills.

The National Trust for Education Collaboration researched the state of education and technology in India, which was published online. The research identified two issues that need immediate resolution in terms of educational attainment and technological development.

- **Restriction of access to innovative technology**

Implementation will be sluggish, and capacity will be restricted as long as a plan and policy are absent. The objectives are unclear, and there is no overarching plan that is implemented on a systemic level. The availability of technology is unevenly spread among regions and quintiles. With no clear, integrated provincial policies, progress is fragmented and driven by solution suppliers rather than by the government. Solutions are not aware of the limitations imposed by available resources.

- **Change management at the systemic level**

According to the report, there is significant opportunity for improvement in seven distinct aspects of the educational system. Of those, change management at the systemic level should be given the utmost attention, and ICT-enabled evaluation should be taken into consideration (Butt 2020).

2.7 OPPORTUNITIES OF THE FOURTH INDUSTRIAL REVOLUTION IN THE EDUCATION SYSTEM

- **Conducive to innovation and creating thinking**

Universities must foster an atmosphere conducive to innovation and creative thinking. This is a response to India's history and the political structure on the other side of the world. There has been significant transformation in schooling and the whole education system – not just in the United States but also across the world. The 4IR provides educational institutions in India with an opportunity to foster a culture of creativity and invention (Figure 2.4). Approaches for the 4IR science, technology, engineering, and mathematics (STEM) fields should be updated on a timely basis (Scepanovic 2019).

- **Incorporate Fourth Industrial Revolution solutions into the new curriculum**

The new curriculum should emphasize 4IR collaboration abilities in addition to the need for 4IR literacy. Aside from that, 4IR education solutions need institutional reform in order to accommodate the development of new interdisciplinary scientific programs.

FIGURE 2.4 Opportunities of the Fourth Industrial Revolution in the education system.

- **Resolution of social exclusion problems**

The use of technology may assist in the resolution of social exclusion problems. In other words, new technical advancements may be utilized to reduce the disparity between the rich and the poor, as well as between races.

- **Develop collaborative relationships with other stakeholders**

Furthermore, the 4IR provides educational institutions with the opportunity to develop collaborative relationships with other stakeholders, especially with government and commercial sector organizations. The Ministry of Human Resource Development of India created a 4IR commission comprised of a diverse group of individuals, including professors from various educational institutions (Ivaldi, Scaratti, and Fregnan 2021). IBM, Microsoft, and other corporate sector players had a significant role in the 4IR's development. These businesses or stakeholders may be able to collaborate on research and development. Furthermore, educational institutions in India have the opportunity to cooperate with educational institutions in other areas of the world (Chweya et al. 2020).

- **Cultivation of Innovative Talents**

In old times, India did not have a system of education that encouraged the development of creative and inventive thinkers. However, now the success of 4IR in teaching in India's education system depends on innovative talent being cultivated. Encouragement of innovation and the development of creating potential are closely related to the success and expansion of research and development. In addition, governments, in collaboration with higher educational institutions, must make significant investments in research and innovative university curriculum in order to get the greatest possible outcomes from new technology advancements.

2.8 RECOMMENDATIONS FOR ENHANCING EDUCATION IN INDIA IN THE LIGHT OF TECHNOLOGICAL ADVANCEMENTS

Following is an in-depth study of suggestions for enhancing education in India in the light of technological advancements. The following conclusions were reached:

- The educational system's capacity to incorporate information and communication technologies (ICTs) into education must be enhanced to guarantee long-term sustainability.
- Instead of focusing on student achievement, educators should instead focus on building learner capacity (Teo et al. 2021).
- There has to be a gradual strategy with realistic intermediate goals. It is important to evaluate progress using a variety of criteria, with the primary emphasis being on assessing the system's ability to incorporate technology.
- Important steps forward include ensuring that the ICT integration system is strategically aligned, strengthening the focus on pedagogy in ICT projects, and improving teachers' skills, confidence, and willingness to integrate ICT into their teaching at all educational levels (Zuiderwijk, Chen, and Salem 2021).

- In order to fulfill the 4IR criteria, institutions must constantly benchmark their management systems, self-awareness, methods, and technology against their peers.

2.9 CONCLUSION

The Fourth Industrial Revolution is here, and it's centered on information and communications technology (ICT), technological advancement, innovation, and creativity. In this study, the characteristics of the 4IR, Big Data, AI, robotics, ICT, 3D printing, and quantum computing were discussed. The Fourth Industrial Revolution requires a new digital approach to education, research and development, and skills development. Many challenges were found when adapting the 4IR, including pedagogical adaptation, teacher development, increased funding and investment in technological development and infrastructure, and increased funding and investment in teacher development. Getting involved in the digital economy and collaborating with others is easy thanks to the 4IR and its numerous possibilities.

REFERENCES

Bjola, Corneliu. 2021. "AI for Development: Implications for Theory and Practice." *Oxford Development Studies* 3 (2): 1–13. https://doi.org/10.1080/13600818.2021.1960960.

Brocal, Francisco, Cristina González, Dragon Komljenovic, Polinpapilinho F. Katina, Miguel A. Sebastián, and Jorge Luis Garciá-Alcaraz. 2019. "Emerging Risk Management in Industry 4.0: An Approach to Improve Organizational and Human Performance in the Complex Systems." *Complexity* 2019: 67–72 https://doi.org/10.1155/2019/2089763.

Butler, Adam. 2018. "The Fourth Industrial Revolution and Education." *South African Journal of Science* 114 (5–6): 17–159. https://doi.org/10.17159/sajs.2018/a0271.

Butt, Javaid. 2020. "A Conceptual Framework to Support Digital Transformation in Manufacturing Using an Integrated Business Process Management Approach." *Designs* 4 (3): 1–39. https://doi .org/10.3390/designs4030017.

Chigwada, Josiline Phiri, and Ngozi Maria Nwaohiri. 2021. *Examining the Impact of Industry 4.0 on Academic Libraries. Examining the Impact of Industry 4.0 on Academic Libraries.* Emrald insight. https://doi.org/10.1108/9781800436565.

Chong, Siewhui, Guan Ting Pan, Jitkai Chin, Pau Loke Show, Thomas Chung Kuang Yang, and Chao Ming Huang. 2018. "Integration of 3D Printing and Industry 4.0 into Engineering Teaching." *Sustainability* 10 (11): 63–69. https://doi.org/10.3390/su10113960.

Chweya, Ruth, Samuel-soma M Ajibade, Abba Kyari Buba, and Moveh Samuel. 2020. "IoT and Big Data Technologies: Opportunities and Challenges for Higher Learning." *International Journal of Recent Technology and Engineering* 9 (2): 909–13. https://doi.org/10.35940/ijrte. b3943.079220.

Collins, Harry. 2021. "The Science of Artificial Intelligence and Its Critics." *Interdisciplinary Science Reviews* 46 (1–2): 53–70. https://doi.org/10.1080/03080188.2020.1840821.

Dash, Sabyasachi, Sushil Kumar Shakyawar, Mohit Sharma, and Sandeep Kaushik. 2019. "Big Data in Healthcare: Management, Analysis and Future Prospects." *Journal of Big Data* 6 (1): 23–27. https://doi.org/10.1186/s40537-019-0217-0.

Elayyan, Shaher. 2021. "Education According to the Fourth Industrial Industrial Revolution." *Journal of Educational Technology and Online Learning* 4 (1): 23–30. https://doi.org/10.31681/ jetol.737193.

Gleason, Nancy W. 2018. "Higher Education in the Era of the Fourth Industrial Revolution" 1–229. Palgrave Macmillan. https://doi.org/10.1007/978-981-13-0194-0.

Gualtieri, Luca, Erwin Rauch, and Renato Vidoni. 2021. "Emerging Research Fields in Safety and Ergonomics in Industrial Collaborative Robotics: A Systematic Literature Review." *Robotics and Computer-Integrated Manufacturing* 67 (July 2019): 101998. https://doi.org/10.1016/j.rcim.2020.101998.

Hoosain, Mohamed Sameer, Babu Sena Paul, and Seeram Ramakrishna. 2020. "The Impact of 4ir Digital Technologies and Circular Thinking on the United Nations Sustainable Development Goals." *Sustainability* 12 (23): 1–16. https://doi.org/10.3390/su122310143.

Howell, Emily, Sara Perez, and W. Todd Abraham. 2021. "Toward a Professional Development Model for Writing as a Digital, Participatory Process." *Reading Research Quarterly* 56 (1): 95–117. https://doi.org/10.1002/rrq.294.

Ivaldi, Silvia, Giuseppe Scaratti, and Ezio Fregnan. 2021. "Dwelling within the Fourth Industrial Revolution: Organizational Learning for New Competences, Processes and Work Cultures." *Journal of Workplace Learning.* 2: 44–52. https://doi.org/10.1108/JWL-07-2020-0127.

Jules, Tavis. 2019. *Conceptualizing Education Governance at the Genesis of the Policy Environment in the Fourth Industrial.* Emerald Group Publishing Limited.

Kocdar, Serpil, Aras Bozkurt, and Tulay Goru Dogan. 2021. "Engineering through Distance Education in the Time of the Fourth Industrial Revolution: Reflections from Three Decades of Peer Reviewed Studies." *Computer Applications in Engineering Education* 29 (4): 931–49. https://doi.org/10.1002/cae.22367.

Lee, Min Hwa, Jin Hyo Joseph Yun, Andreas Pyka, Dong Kyu Won, Fumio Kodama, Giovanni Schiuma, Hang Sik Park, et al. 2018. "How to Respond to the Fourth Industrial Revolution, or the Second Information Technology Revolution? Dynamic New Combinations between Technology, Market, and Society through Open Innovation." *Journal of Open Innovation: Technology, Market, and Complexity* 4 (3): 53–62. https://doi.org/10.3390/joitmc4030021.

Liao, Yongxin, Eduardo Rocha Loures, Fernando Deschamps, Guilherme Brezinski, and André Venâncio. 2018. "The Impact of the Fourth Industrial Revolution: A Cross-Country/Region Comparison." *Production* 28 (11): 2–18. https://doi.org/10.1590/0103-6513.20180061.

Matthews, Matthews, Adam, McLinden, Mike and Greenway, Celia. 2021. "Rising to the Pedagogical Challenges of the Fourth Industrial Age in the University of the Future: An Integrated Model of Scholarship." *Higher Education Pedagogies* 6 (1): 1–21. https://doi.org/10.1080/23752696.2020.1866440.

Mohajan, Haradhan. 2019. "The First Industrial Revolution: Creation of a New Global Human Era." *Journal of Social Sciences and Humanities* 5 (4): 377–87.

Nagy, Judit, Judit Oláh, Edina Erdei, Domicián Máté, and József Popp. 2018. "The Role and Impact of Industry 4.0 and the Internet of Things on the Business Strategy of the Value Chain-the Case of Hungary." *Sustainability* 10 (10): 43–58. https://doi.org/10.3390/su10103491.

Nakicenovic, Nebojsa, Dirk Messner, Caroline Zimm, Geoff Clarke, Johan Rockström, Ana Paula Aguiar, Benigna Boza-Kiss, et al. 2019. "*The Digital Revoluion and Sustainable Development: Opportunities and Challenges Report.*" https://doi.org/10.22022/TNT/05-2019.15913.

Oke, Adekunle, and Fatima Araujo Pereira Fernandes. 2020. "Innovations in Teaching and Learning: Exploring the Perceptions of the Education Sector on the 4th Industrial Revolution (4IR)." *Journal of Open Innovation: Technology, Market, and Complexity* 6 (2). https://doi.org/10.3390/JOITMC6020031.

Rapanyane, Makhura B., and Florence R. Sethole. 2020. "The Rise of Artificial Intelligence and Robots in the 4th Industrial Revolution: Implications for Future South African Job Creation." *Contemporary Social Science* 15 (4): 489–501. https://doi.org/10.1080/21582041.2020.1806346.

Rotatori, Denise, Eun Jeong Lee, and Sheryl Sleeva. 2021. "The Evolution of the Workforce during the Fourth Industrial Revolution." *Human Resource Development International* 24 (1): 92–103. https://doi.org/10.1080/13678868.2020.1767453.

Scepanovic, Snezana. 2019. "The Fourth Industrial Revolution and Education." 2019 8th Mediterranean Conference on Embedded Computing, MECO 2019 - Proceedings 0 (June): 8–11. https://doi.org/10.1109/MECO.2019.8760114.

Schwab, Klaus. 2016. *The Fourth Industrial Revolution*. World Economic Forum.

Sima, Violeta, Ileana Georgiana Gheorghe, Jonel Subić, and Dumitru Nancu. 2020. "Influences of the Industry 4.0 Revolution on the Human Capital Development and Consumer Behavior: A Systematic Review." *Sustainability* 12 (10): 33–36. https://doi.org/10.3390/SU12104035.

Siponen, Mikko, and Richard Baskerville. 2018. "Intervention Effect Rates as a Path to Research Relevance: Information Systems Security Example." *Journal of the Association for Information Systems* 19 (4): 247–65. https://doi.org/10.17705/1jais.00491.

Sivarajah, Uthayasankar, Muhammad Mustafa Kamal, Zahir Irani, and Vishanth Weerakkody. 2017. "Critical Analysis of Big Data Challenges and Analytical Methods." *Journal of Business Research* 70: 263–86. https://doi.org/10.1016/j.jbusres.2016.08.001.

Spöttl, Georg, and Lars Windelband. 2021. "The 4th Industrial Revolution: Its Impact on Vocational Skills." *Journal of Education and Work* 34 (1): 29–52. https://doi.org/10.1080/13639080.2020.1858230.

Sutherland, Ewan. 2020. "The Fourth Industrial Revolution: The Case of South Africa." *Politikon* 47 (2): 233–52. https://doi.org/10.1080/02589346.2019.1696003.

Teitel, Simón. 2000. "Manufacturing Industry." *Technology and Skills in Zimbabwe's Manufacturing* 2: 22–29 https://doi.org/10.1057/9780230514027_2.

Teo, Timothy, Siobhan Unwin, Ronny Scherer, and Veronica Gardiner. 2021. "Initial Teacher Training for Twenty-First Century Skills in the Fourth Industrial Revolution (IR 4.0): A Scoping Review." *Computers and Education* 170 (May): 104223. https://doi.org/10.1016/j.compedu.2021.104223.

Tsiligiris, Vangelis, and Dorothea Bowyer. 2021. "Exploring the Impact of 4IR on Skills and Personal Qualities for Future Accountants: A Proposed Conceptual Framework for University Accounting Education." *Accounting Education* 4 (1): 1–29. https://doi.org/10.1080/09639284.2021.1938616.

Zuiderwijk, Anneke, Yu Che Chen, and Fadi Salem. 2021. "Implications of the Use of Artificial Intelligence in Public Governance: A Systematic Literature Review and a Research Agenda." *Government Information Quarterly* 38 (3): 101577. https://doi.org/10.1016/j.giq.2021.101577.

CHAPTER 3

Integration of IoT and Big Data Technologies for Higher Education

Amrinder Singh and Geetika Madaan

CONTENTS

DOI: 10.1201/9781003254942-3

3.1 INTRODUCTION

In addition to the many creative implementations in every area of businesses and society, Internet of Things (IoT) and Big Data have had a major impact on the digital transformation process (Sima et al., 2020). Furthermore, research shows that the rise of innovation and the internet has given the education industry access to teaching, learning, and assessment strategies that are more diverse in terms of where they are used and how they are evaluated (Coman et al., 2020). With the development of social media, IoT, and multimedia data – whether structured or unstructured – businesses are producing more data (Sarker, 2021). Online and distant learning are now more accessible than ever before, thanks to digital innovation and social media, which allows for universal and broad access at a low cost for everyone (Muthuprasad et al., 2021). Increasing the amount of data collected through the IoT has made a major contribution to the Big Data environment. Due to the processing and gathering of data through different sensors in the IoT ecosystem, the growth of the IoT has generated difficulties in Big Data analytics (Muthuprasad et al., 2021). According to studies, the IoT and Big Data have a two-way relationship. Thus, although IoT is a major generator of Big Data, it is also a substantial contributor to Big Data analytics, which is critical for the improvement of IoT processes and services (Motlagh et al., 2020). 'Big data' refers to an enormous quantity of information that requires contemporary technological advances in order to extract useful information. In the same way, the IoT is an internet device that involves connecting different gadgets through certain intelligent connections in order to share information. Studies predict that by 2019, the Big Data business would be worth over US$125 billion. The IoT is also expected to increase at a compound annual rate of 20% by 2020, reaching up to 7.1 trillion dollars, with sensory devices expected to reach 190.6 billion dollars by 2021. IoT data, with its heterogeneity, variety, and unstructured data, will overtake traditional Big Data as the most significant source of Big Data once the number of sensors hits a trillion (Krishnamurthi et al., 2020).

IoT will provide context data to Big Data systems since it aims to connect 'sensors and other devices' to the Internet. This is presently a challenge for Big Data systems. Integrating Big Data with IoT also means that Big Data will help IoT data be processed, resulting in new IoT solutions being developed (Krishnamurthi et al., 2020). In the IoT, research has shown that Big Data has the following characteristics: Many terminals produce a huge amount of data and details become critical following thorough examination of the semi-structured and unstructured information provided (Khare and Totaro, 2019). A more convenient method to access the internet is provided by the IoT, which enables any device to communicate with other IoT devices from anywhere at any time. The current Big Data boom, on the other hand, has sparked a rapid digital change in society and spread throughout the world with many profitable applications (Gulia and Chahal, 2020). Big data and the IoT have, on the other hand, made major contributions to digital transformation via the development of many innovative applications in almost every sector of business and society (Gulia and Chahal, 2020). Combining Big Data and IoT technology offers up a plethora of possibilities for not only smart cities, but also for academic institutions looking to innovate. Several studies have been done on Big Data, as well as on IoT technology.

Higher education institutions are not following a trend when it comes to incorporating new information technology into their practices. Even while traditions are essential in higher education institutions, there are at least two reasons that make change necessary (Darling-Hammond et al., 2020): the first and most critical is that the world is undergoing rapid transformation; everyone is involved in the process, from students to business people to academics, and must adapt quickly; the second argument is that new options may improve the effectiveness of education, even when it is done within the conventional paradigm (Kromydas, 2017). Undoubtedly, modernizing university curricula is necessary if students are to be prepared for the increasingly digital and automated world of Industry 4.0 (Rapanta et al., 2020). According to the definition, Industry 4.0 calls for the use of a network strategy that is based on the ability to create intelligent systems. A network strategy that relies on the ability to build intelligent systems is required by Industry 4.0 (Yang and Gu, 2021).

In addition, products and components are covered in this category. In the context of Industry 4.0, where the framework is still being developed, 'smart' education would be a logical step. However, preparing education for this digital revolution is a significant challenge (Oztemel and Gursev, 2020). Students must be ready to meet the new requirements that may arise because of the Fourth Industrial Revolution while considering their future career fields.

The present study is based on one of today's most pressing issues: how can information technology innovations utilizing technologies like Big Data and the IoT be integrated with higher education? It goes without saying that there are many methods, as well as an almost limitless amount of information that may be helpful in the preparation of such an action.

Some important points will be made, and real-world examples will be used to support them. The primary focus of the project is on the storage and in educational contexts, and huge volumes of data processing, with a particular emphasis on the use of information technology (IT) in certain procedures (Wynn and Clarkson, 2018). Unquestionably, a university stands to gain from the collection of useful information.

Higher education's social purpose is intertwined with the new IT's importance. Universities must continue to act as testing grounds for educating the next generation and inventing in order to meet the needs of the digital revolution (Sun and Chen, 2016). Education 4.0 emphasizes close cooperation with the social environment, businesses, and other stakeholders. For universities to accept know-how and new ideas, they must change their administrative paradigms, invest in people, and put the interests of society above all else (Camilleri, 2017).

The parts of this chapter are structured as follows: An introduction of Big Data and IoT technologies, with a focus on their application in higher education. How in India has implemented remote learning in higher education using contemporary IT is also discussed in this part. This section, which is divided into many subsections, discusses the implementation of strategies in a few unique situations. In the last part, the authors address future research possibilities and offer and practical applications of their findings. Although basic thoughts regarding the new methodology's effect on the educational process have been presented, the chapter does not purport to provide a comprehensive didactical study of it.

There is some evidence to support the idea that adopting IoT and Big Data integration solutions may alleviate concerns about data management and storage. This will be useful for a variety of educational purposes, such as predicting student success. With these advances, massive amounts of data can be collected and processed in real time without human intervention. However, educational institutions are hampered by the challenge of effectively and efficiently managing and analyzing the information generated by educational resources. Older methods can only handle a limited number of data applications because of the large quantity of collected data. This resulted in the utilization of Big Data applications for processing their educational data. Big Data and the IoT provide both possibilities and problems for academic institutions (Hassan, 2020). This paper is structured as follows: Part II discusses the IoT and the use of Big Data. Using IoT and Big Data, Part III examines the educational implications of IoT. Part IV examines how IoT and Big Data impact schooling. Conclusions are discussed in Part V.

3.2 CONCEPTUAL FRAMEWORK FOR THE IOT AND BIG DATA

3.2.1 IoT and Big Data

3.2.1.1 IoT

The IoT gives businesses a foundation on which sensory gadgets and devices may seamlessly connect, allowing information to travel between locations with ease. Studies have shown that IoT may allow for self-configuration of standards and protocols. IoT communication devices are often integrated with sensor devices in the real world (Sethi and Sarangi, 2017). Because of this, IoT provides an upgrade to current advances with many devices in the surrounding area. Because of artificial intelligence, institutions are able to make well-informed decisions that improve student learning experiences, operational efficiency, and the security of the institution (Colakovic and Hadzialic, 2018).

As a result, the IoT provides ongoing assistance in higher education, opening new career possibilities that are made possible by ubiquitous computers and technology.

3.2.1.1.1 IoT in Education: State of the Art and Future Prospects The IoT is a cutting-edge technology that links devices and equipment all around the globe into one global network so they may interact with one another (Motlagh et al., 2020). We want to link all gadgets to an internet-controllable network that will supply us with data and information in real time while also allowing us to engage with those who use it in that time. IoT was originally used to describe devices connected through radio frequency identification (RFID) in the late 1990s when it first came on the scene. The IoT is a global network of interconnected digital devices, with an ever-increasing number of devices participating (Perwej et al., 2019). Consequently, civilization is undergoing a technological revolution that touches on everything from the interaction between objects and daily activities to the most complex procedures for building whole communities (Figure 3.1).

3.2.1.2 Big Data

Big Data refers to huge amounts of data that may be either organized or unstructured in nature. The information is gathered and evaluated in order to be utilized in educational

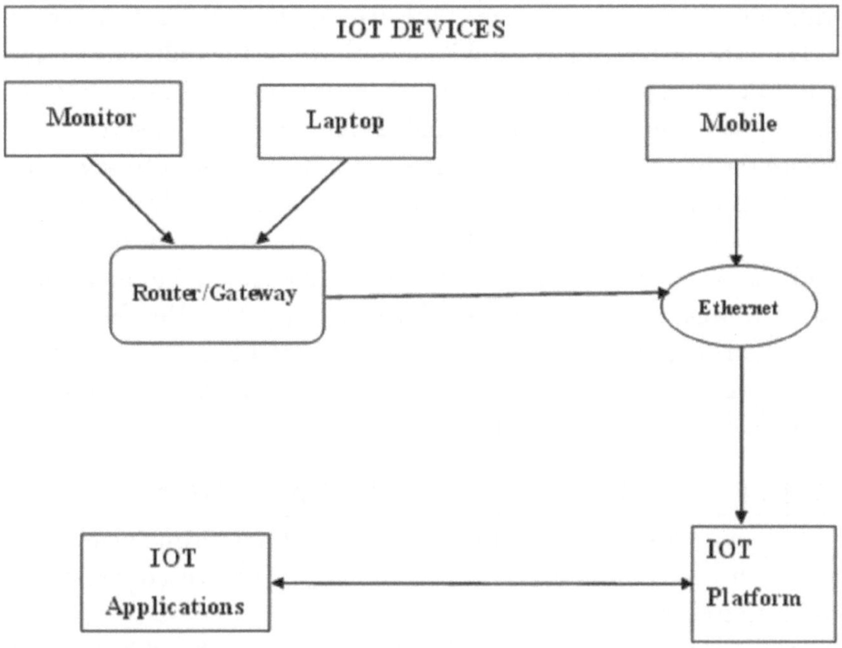

FIGURE 3.1 Typical organization of an IoT system.

institutions (Perwejet al., 2019). Big data has a major effect on instructors, school structures, students, and curriculum in the field of educational technology (Bamiah, Brohi and Rad, 2018). Furthermore, analyzing large amounts of data may aid in the identification of pupils at risk, the verification that learners are making the necessary progress, and the creation of a better system to assess and assist instructors and school administrators (Huber and Helm, 2020).

The IoT offers businesses with new and creative methods to control and monitor distant activities. From a conceptual standpoint, it provides the opportunity to link the physical world with the digital world through the Internet (Attaran, 2017). 'A dynamic worldwide network infrastructure with self-configuring capabilities based on standard protocols', according to a generally recognized description of the contemporary IoT and smart communication protocols in which real and virtual objects are merged smoothly into the information network while retaining their own identities.

The IoT has resulted in massive amounts of data being generated by billions of linked devices (Attaran, 2017). The creation of a new IoT has opened previously unimaginable possibilities for harnessing information's power (Alam, 2018). With the enormous quantity of data produced by industrial sensors, it is possible to make industrial production processes more efficient, while data from smart devices may be used to enhance a variety of human experiences, from home environment monitoring to specialized social network content (Nizetic et al., 2020). As previously stated, the method is strongly tied to Big Data technology, which will be examined in detail below.

IoT technology will be used in education, especially higher education, according to the author's research. There are many facets to the path. As a result, it is naturally a hot topic in engineering education right now. One of the main topics of this article is higher education's use of IoT technologies (Ramya and Poongodi, 2021). The route has many aspects.

And it is a popular subject in engineering schools right now. As we'll see later, it's also a critical part of many forms of electronic and remote learning. However, as time passes, it blends in with the rest of what you study in school. Because of IoT, conventional educational and learning techniques are being restructured more efficiently than before. Changes to the Learning Management System (LMS) in general are likely to be the most significant (Mershad and Wakim, 2018). Flexibility and dynamic changes in the curriculum allow the system to gain self-organizational abilities (Aldowah et al., 2017). This means that integrating IoT devices with a knowledge base has the potential to have a significant influence on how education develops going forward. While growth is possible, it is not only dependent on the acquisition of new technological skills.

3.3 BIG DATA IN HIGHER EDUCATION: A NEW APPROACH

More and more educational institutions are using Big Data technologies, and although there are some overviews available, their application is fraught with difficulties. Every day, more data is being collected.

In the year 2000, the quantity of digital information represented just a quarter of the entire amount of information available across the globe (Motlagh et al., 2020). There is now a zettabyte (ZB) of information saved across the globe, with non-digital material accounting for less than 2% of the total stored information. Internet users generate, on average, more than 2.5 quintillion bytes of data generated every day. Large-scale data creation necessitates the development of Big Data tools together with the IoT and storage solutions for Big Data (Coman et al., 2020).

Aside from that, getting information is tough due to the variety of delivery methods. Except for data that has been organized (for example, data from sensors and log files, financial information, input data, and information about visitors), scientific knowledge data, images or videos, and meteorological data are examples of unstructured large data, and there are many different types of data that may be gathered from satellites as well (Motlaghet al., 2020). These include oceanographic and seismographic observation data as well as satellite images.

This enormous quantity of data may be handled efficiently using methods such as real-time and parallel processing (Kumar, Tiwari and Zymbler, 2019). The employment of both real-time and parallel methods is beneficial for addressing difficult problems. To manage and extract hidden value and information from all the Big Data, new architectures, techniques, algorithms, and analytics must be developed.

Techniques, methods, and processes for data processing with 'Big Data' – we're talking about managing enormous quantities of organized and unstructured data in a way that makes sense to the public (Muthuprasad et al., 2021). Increasing productivity, creating new goods, and making the market more competitive are all goals which technology serves. The use of Big Data analytics may also help governments and businesses make crucial decisions and improve policy making in the direction of economic growth.

It's possible to think about Big Data in this way instead of just in terms of technology. Big data encompasses a broad variety of information, as we've witnessed increases in both the amount and speed of information, according to the Oracle website. Big Data applications in education are projected to be included in research as a starting point. Some positive tendencies are emphasized, but there are some negative ones, so keep that in mind while making changes. Students and teachers need time to prepare for the changes, and this time must be provided to them (Kromydas, 2017).

The following five characteristics may be applied to large amounts of data:

1. *Capacity:* Big Data generally refers to a very large amount of data, often measured in zettabytes (ZB).
2. *Change:* It contains information that is a mix of structured, unstructured, and semi-structured data gathered from a wide range of sources such as RFID, web searches, social media platforms like Facebook and Twitter, mobile devices with sensors such as GPS and accelerometers, industrial sensors with high levels of fidelity, and live streaming video.
3. *Swiftness:* Large amounts of Big Data are generated at different rates, spanning from milliseconds to days to years, with variable processing needs.
4. *Charge:* Data is deemed valuable if relevant information (from a commercial or technical viewpoint) can be retrieved from huge datasets when individual data points may have little value on their own, according to some academics.
5. *Reliability:* This is a reference to the data's degree of correctness and reliability. Because of worries about privacy and security, a significant percentage of IoT users may be hesitant to submit accurate data.

It's crucial to be able to analyze and integrate data collected daily while utilizing Big Data, regardless of where it occurs and maintained during everyday data management in repositories within institutions (Bamiah, Brohi and Rad, 2018). Big Data management must also provide reliable protection for private information if it is to be used with this technology.

For educational purposes, Big Data have created an entirely new area of study with an emphasis on the analysis of educational records (Oztemel and Gursev, 2020). The advances in Big Data technology have undoubtedly aided in the promotion of analytics in higher education. Big data analytics may be valuable to universities since it allows them to keep tabs on their students' academic progress and needs (Darling-Hammond et al., 2020).

Due to the fast development of Big Data analytics, universities must align their management and measures with it as well. In the literature study, it was found that new information technologies like Big Data and IoT have a research gap that must be addressed before they can be effectively used in the educational sectors. As key facilitators for IoT applications, large data storage, processing, and analytics are discussed using a converging teaching platform.

3.4 BENEFITS OF INTEGRATING IOT AND BIG DATA IN EDUCATION

A variety of intelligent technologies are helping to make smart education a reality. These include things such as learning analytics, Big Data, cloud computing, and the IoT (Sethi and Sarangi, 2017). According to the findings of this study, IoT data are different from global Big Data. Following that, we'll talk about the educational advantages of IoT and Big Data. One of the most important roles that IoT and Big Data will play in education is delivering customized content directly to students' desks. IoT and Big Data have been linked in research, it turns out. This is due to the fact that IoT on smart devices generates a lot of data, and that data must be evaluated using Big Data analytics. In order to enhance the various processes of IoT devices, Big Data itself demands numerous types of analytics. As a result, combining IoT and Big Data in several fields, such as current education structures, intelligent transportation, environmental protection, and agriculture, would result in improved operations (Krishnamurthi et al., 2020). Education's primary purpose is to enhance student learning, boost efficiency, and offer a teaching and training environment that is essential and beneficial for students. Big data technology may make this easier by allowing students, instructors, and administrators to communicate and be available at the same time.

This is made feasible by data mining techniques used in model building. Data mining algorithms. The models will build adaptive learning structures based on the model's predictions, with the potential to change the experience of learners or assist management with recommendations for student help (Gulia and Chahal, 2020). IoT may also help students study better. With IoT, systems become more effective and flexible, and new business possibilities and income streams are created. For example, Columbia University research found that incorporating IoT improved student performance. Furthermore, the IoT facilitates and improves education quality. As a result, IoT has the potential to enhance, assist, and promote lifelong learning (Khare and Totaro, 2019).

Greater degrees of competitiveness, accreditation, assessments, and restrictions are being implemented in the field of higher education. The use of new digital learning frameworks is also being used by educational institutions to help identify students' behaviors and their conduct. As a result, decision-makers are required to come to decisions based on pertinent information derived from the data under investigation. The goal is to determine, aside from academic advancement, the rates of success, patterns of achievement, and challenges experienced by students. Big data technologies, as a result, are concerned with enormous amounts of educational data that are used for assessing and tracking students' previous knowledge, keeping track of all the activities and inputs that students make over the course of their education (Wynn and Clarkson, 2018). Smart devices, on the other hand, have the potential to be important in teaching and learning since they have the potential to enhance the learning process. As a result, management will be better able to choose appropriate practices and simplify the learning processes. In this way, Big Data may aid to enhance teaching and learning while also assisting an institution in achieving academic goals. The research concludes that both 'Big Data and the IoT (IoT)' are very important in learning and teaching (Coman et al., 2020). A successful institution's accomplishment rate, shortcomings and strengths, and position in relation to other institutions must be traced and established. Analysis of the company condition, academics accreditation, and future planning and management of their strategy are all made possible by Big Data technology.

According to the findings of the research, higher education institutions may benefit from the use of Big Data to provide unparalleled visualization to stakeholders for seamless management and encouraging results (Mershad and Wakim, 2018). This offers proof to accomplishment rates and helps to identify areas for development to be more proactive.

Educational Institutions can more easily gather data from sensors and wearables thanks to the IoT, which gives them the ability to act on that data. For example, the use of fast tracking learners may be aided with response codes, radio frequency chips, and feedback (Coman et al., 2020). However, most of the information collected in higher education is not properly used due to a lack of appropriate technology to convert unstructured information into relevant data, as is the case in many other fields. As a result, data analytics is available to uncover this process for the purpose of furthering the growth of an institution using Big Data (Darling-Hammond et al., 2020). Learner behavior, teacher progress, and course results are all tracked and analyzed using Big Data technology, which provides decision-makers with the information they need for detection, understanding, analysis, and forecasting. This allows teachers to assess students' knowledge and choose the most effective learning techniques for everyone. Thus, it provides teachers with opportunities to learn about the most up-to-date techniques and processes relevant to their areas of teaching expertise. Similarly, using the IoT, various tasks may be carried out, such as tracking learners, collecting information to determine which pupils need more attention, and assisting teachers in altering designs and processes for future courses. According to the findings of the research, this involves identifying learner behavior and accomplishments to help in the provision of resources when necessary. When it comes to higher education, Big Data and the IoT play a crucial role (Darling-Hammond et al., 2020).

An educator may utilize Big Data to gain fast, actionable feedback on how their course and teaching and assessment procedures are doing, along with the tactics they are utilizing.

As a result of the processes' monitoring, it is possible to identify student weaknesses and failure risk early on, depending on the learner's ability and knowledge level. The implications of this include that Big Data may offer a more in-depth identification of students' study needs or learning problems (Alam, 2018). In a similar vein, the IoT, like Big Data, enables effective and attentive administration of every device that is used, due to the belief that proper monitoring of students, staff, and resources through IoT deployment would lead to the desired end result for learners. As a result, having a thorough understanding of their surroundings will assist them in providing essential information when it is needed to further enhance the learning process. Institutions are exploring ways to increase numbers at a lower cost while yet maintaining affordability. Research shows that Big Data may not only help students succeed, but it can also help institutions and learners save money (Aldowah et al., 2017). To enhance decision-making, it has been proven that Big Data may be used at a low cost. In a similar vein, the IoT has been touted as a cost-saving tool for organizations. The IoT has a beneficial impact on educational models that make use of sensors, thus lowering the overall learning costs.

3.5 CHALLENGES OF USING IOT AND BIG DATA IN EDUCATION

No matter how many universities have started to offer different levels of data science degrees, the problem remains: traditional academic curriculum is still lagging behind Big Data advances (Nizetic et al., 2020). Big Data integration into data science education is one of the difficulties, using data science to teach students how to create tools and extract information from large datasets, expanding the curriculum to include modern topics like intelligent enhancement and learning about cognitive computing and data management as part of a general education programs. Disaster recovery plans and strong passwords fail to defend the IoT, Big Data, and even cloud computing against cyberattacks because they don't safeguard high-tech advancements from cyberattacks (Attaran, 2017). Educational institutions require adequate standards to govern and monitor intellectual property and access to information in addition to safeguarding the storage, transmission, and processing of huge volumes of varied instructional data types. Researchers have also emphasized the importance of security on the IoT as a problem. For example, according to research conducted by the author, security is a concern in using IoT in college and university classrooms.

With Big Data and IoT, transparency is essential for demonstrating student identification and making informed decisions. The learner's data is protected under the terms of privacy (Yang and Gu, 2021). However, there are certain instances in which the learners fail to give the information needed. The potential of a malevolent individual using learner information (educational and financial records, performance, and other activities) exists in the Big Data environment. The right to privacy of a student may be violated if such information is necessary for predicting academic success and future repercussions. The potential of educational breaches exists, as shown by the breach of 14.8 million data from educational institutions that occurred in 2021. The problem of privacy has not been adequately addressed with the IoT. For example, privacy issues are a concern in the adoption of new technologies such as the IoT. In addition, according to other research, the IoT has a positive impact on education since it improves learning despite the privacy concerns. Many institutions of higher learning do not have established ethical review procedures. Identifying methods for the preservation of private data, personal permission, data ownership, and transparency are among the ethical issues that arise because of this. The use of different devices and methods for data collection and analysis of educational facts raises ethical concerns about the validity of the information being collected. Big data and the IoT applications are generating colossal quantities of data in the educational sector, which is being used.

3.6 BIG DATA ANALYTICS AS A COMPONENT OF THE QUALITY ASSURANCE SYSTEM AT COLLEGES AND UNIVERSITIES

In the educational field, analytics based on Big Data, especially in higher learning, is now presenting a major issue, as previously stated. Educational analytics' strategic objectives are as follows: mentoring, curriculum and learning process creation, as well as support for school administration, are among our offerings. After this transition, institutions began to change their management methods, concentrating more on 'high involvement systems'. The essential principles for utilizing analytics include accountability, openness, and permission, as well as privacy, validity, access, and limiting negative consequences, etc.

There are five major factors for the educational process:

- Coordinating educational institutions' results, aims, and objectives (planned targets) with the current employment trend and the requirements of society.
- Make detailed information on the specialties and circumstances of education accessible to consumers.
- Educational process flexibility to meet changing consumer expectations and needs as well as educational process innovation.
- Match between student abilities and teacher abilities and opportunities.
- Universally accepted criteria for all participants in the educative process.
- Based on success criteria, an aggregated and implemented (detailed) assessment system should be utilized in the monitoring of product quality.
- When analyzing a unit's performance against that of others in dynamics, there are three typical methods (against the backdrop of similar units' previous states), evaluating the unit's existing features considering the desired outcomes.
- Using three sets of indicators to evaluate the quality of education, the management method is a possible alternative way to evaluating educational quality:
- Competence (cost-cutting measures),
- Efficacy (achieving one's goals and ambitions),
- Proficiency (optimal use of available resources while maintaining high performance).
- To evaluate the effectiveness of the methods used for data gathering and processing, and to make choices based on the data gathered, each department has its own standards.
- To reconsider higher education's economic model, and to gather information regarding educational achievements from many sources, the 2013 OECD Report may be the foundation (Figure 3.2).

3.7 HIGHER EDUCATION AND THE USE OF BIG DATA MODELING

Using analytics for learning, we can better understand how people learn as well as the external settings in which they learn it. Being an exceedingly complicated process with many levels and factors, higher education necessitates the use of analytical monitoring (and methods for doing so).

Process and workflow optimization is often achieved via the use of software and analytics training methodologies, to improve an organization's overall performance and to analyze academic and institutional data. Business–academia partnerships are becoming increasingly frequent. To attract and promote these types of partnerships, a university's ability to conduct applied research, disseminate knowledge, and commercialize its findings depends on its ability to use and create innovative technologies.

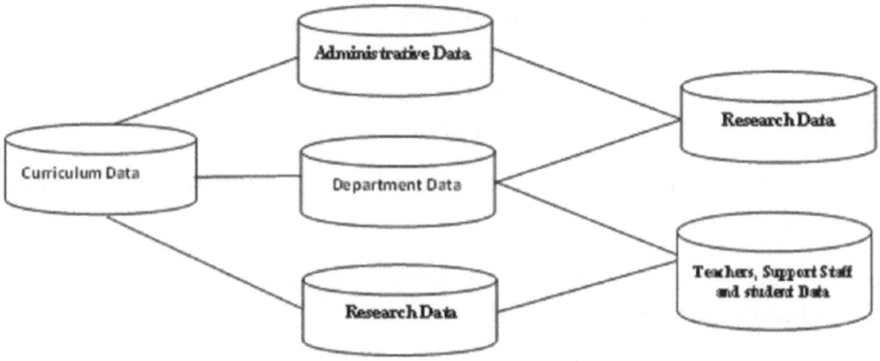

FIGURE 3.2 Education institution adopts Big Data architecture.

Learning and teaching have undergone dramatic transformations because of the digital revolution brought on by new technological advances in the past several decades. A Mobigital virtual environment is being created where individuals may study and teach digitally at any time, from anywhere. Massive Open Online Courses (MOOCs) have emerged due to efforts to find answers to the issue of training classes available to everyone, regardless of income or location, regardless of where one lives or whether one is enrolled as a student. There is broad acceptance of MOOCs across the globe, and this has demonstrated that the teaching method needs a study of their methodological basis to include components in Russian universities in intramural and distant learning processes.

One of the most recent forms of distance learning is MOOCs, which are massive open online courses. Many MOOC-based online learning platforms are seeing an increase in the number of students enrolling in every year and in every course. MOOC course content creation expenses have been reduced through the use of cloud computing and Big Data, making it possible for MOOC providers to make their material freely accessible to learners located anywhere with internet connection.

When it comes to institutional effectiveness and change adoption, it's well-known that an examination of relevant data is the foundation for organizational learning. Institutions may now access previously impossible levels of complexity, speed, and precision thanks to contemporary technology. A result of the growing use of technology in higher education is that students, computer programs, and computer-based systems generate information that is valuable.

3.8 USING THE IOT AND BIG DATA TO CREATE A DIGITAL EDUCATIONAL PLATFORM

3.8.1 Fundamental Requirements for Big Data Deployment

3.8.1.1 Norms for Contemporary Data Gathering and Processing Systems

The IoT paradigm collects data from a multitude of sources, including the internet, social media, and other electronic devices, mobile sensors, RFID, and other technologies. The

Big Data platform is equipped with the necessary technologies to deal with data on its own (structured and unstructured), in addition to data in motion (strong data streams from any sort of source) (Gil et al., 2016). It is necessary to do stream processing: in data storage, there is a way to overcome the hurdles of dimension.

3.8.1.2 Real-Time System Training is Required for the Platform

The aim of training information technology systems is unquestionably the development of their features in two fundamentals aspects: IT awareness and intelligence (sometimes known as 'IT intelligence') are two factors that may impact the quality of a solution. The use of knowledge gained in the past to make decisions in the present or foresee the future is at the heart of many approaches for teaching people and systems today (Khare and Totaro, 2019). Through the use of machine learning technologies, the processed data is mined for important information that can then be presented and used for predictive analysis.

3.8.1.3 Data Management

MapReduce and NoSQL are two strong Big Data technologies that are used to efficiently collect data from diverse sources and process it in accordance with application requirements (Oussous et al., 2018).

3.8.1.4 Information Creation and Search that Are Completely Unrestricted and Adaptable

It is possible for teams of analysts (from IT and business) to conduct free creative searches throughout the whole internet using the toolset of search platforms that have been technologically deployed (Sethi and Sarangi, 2017). A creative team controls modern platforms, which apply the idea of open cyberspace search with input on this team.

Figure 3.4 is an example of an IoT-based Big Data infrastructure.

For the most part, unstructured data is collected from different sources and supplied through stream processing, where they are processed before being organized as the data moves through the architecture (as shown in Figure 3.1).

3.8.2 The Many Distinct IoT Platforms

IoT Platform on Amazon Web Services – connecting to the platform enables communication between IoT devices.

The Bosch IoT Suite – offers IoT device management and analysis capabilities as well as the ability to restrict access to IoT devices to authorized individuals.

IoT by Google – while the focus is on the analytical component, it may be utilized to manage IoT devices.

IBM's IoT platform on the cloud – Various devices and gateways may be linked through the platform, which allows for device management, storage, and access.

Platform for the IoT hosted on Microsoft Azure – is a set of tools for building IoT applications. IoT implementation challenges.

3.8.3 Architecture for IoT and Big Data (Figures 3.3 and 3.4)

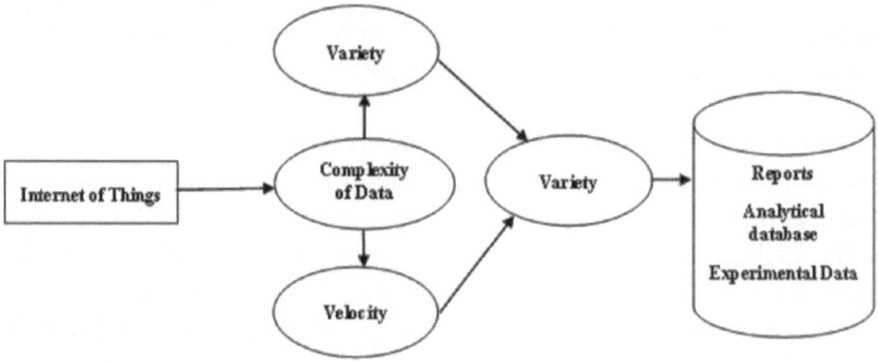

FIGURE 3.3 Architecture for IoT and Big Data.

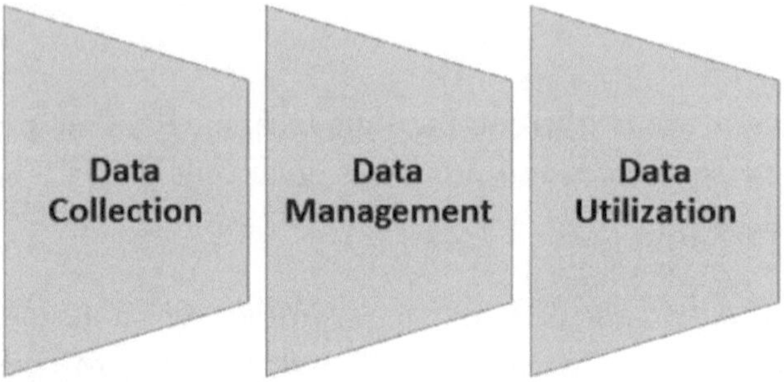

FIGURE 3.4 IoT and Big Data.

3.8.4 Kinds of Cloud Computing

With so many distinct designs, hardware platforms, and protocols in IoT devices, maintaining software becomes more difficult (Khare and Totaro, 2019). This makes it difficult to create a single software solution that can be used by many devices, each with a different architecture (see Figure 3.5).

There are many advantages to use IoT technologies in educational institutions:

- IoT technologies allow for increased efficiency and automation in the educational process.
- Processing huge amounts of data in a unique way.
- Interoperability with other systems made simple by industry standards.

Many academics have addressed the difficulties that large data tools have in terms of both technological and architectural design. Additionally, these authors point out that that emerges from combining Big Data tools with IoT applications with its own set of challenges, including the need for specific industry expertise (Noura, Atiquzzaman and Gaedke, 2019).

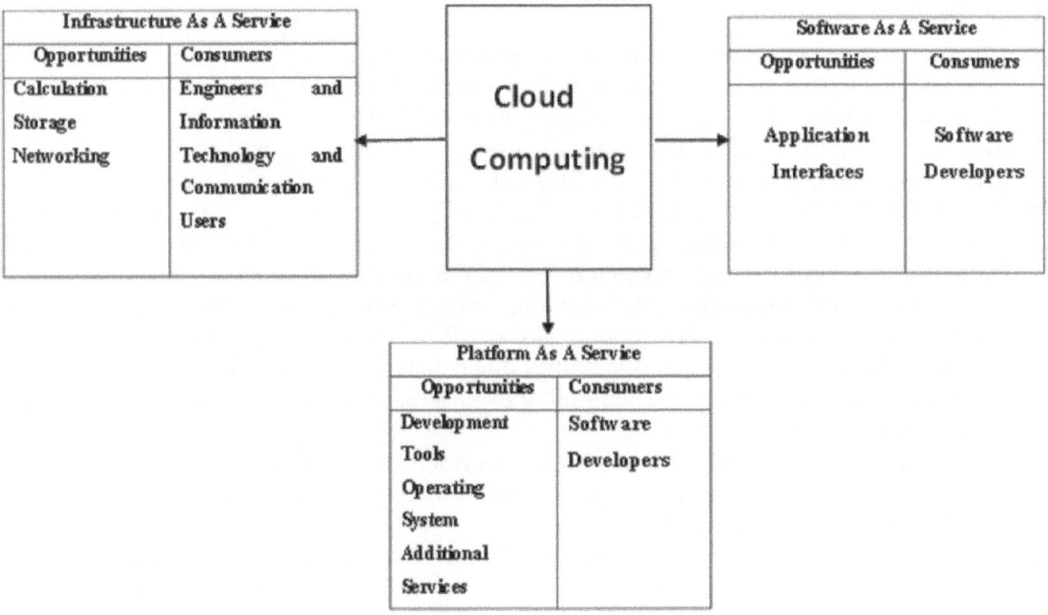

FIGURE 3.5 Cloud Computing.

3.9 CONCLUSIONS

The education industry is seen as a natural place to put IoT and Big Data to use. This paper's primary goal is to describe the many applications for IoT and Big Data in higher education institutions. Additionally, it reveals the challenges associated with using two different types of technology. There have been many studies looking at the need of incorporating IoT and Big Data into the classroom environment to improve both education and learning. The acceptability and potential application of the technologies is still limited, particularly in impoverished nations, even with various IoT and Big Data technologies used in education. It was discovered in the study that it is possible that the IoT and Big Data may enhance individualized learning for students, the learning environment for students, the wide distribution of knowledge, and the academic success of institutions. It is possible for them to help stakeholders in the efficient administration of the project, to track students' movements, and to enhance decision-making. The lack of established ethical review processes in many educational institutions is a weakness for the preservation of individual information and openness, according to the research. This study recommends that universities put their complete faith in IoT and Big Data technologies and reevaluate how they use such advances to assist those universities. It is also necessary to assess the contribution of IoT and Big Data on student performance considering the reviewed research. In addition, issues affecting the use of IoT and Big Data in emerging countries must be understood in higher education. This will have a beneficial impact on how the internet of Things and Big Data are used in higher education.

REFERENCES

Alam, T. (2018) 'A reliable communication framework and its use in Internet of things (IoT)', *International Journal of Scientific Research in Computer Science, Engineering and Information Technology*, 5(June): 23–29. doi: 10.36227/techrxiv.12657158.

Aldowah, H. et al. (2017) 'Internet of things in higher education: A study on future learning', *Journal of Physics: Conference Series*, 892(1). doi: 10.1088/1742-6596/892/1/012017.

Attaran, M. (2017) 'The internet of things: Limitless opportunities for business and society', *Journal of Strategic Innovation and Sustainability*, 12(1), pp. 10–29. Available at: https://ucd.idm.oclc.org/login?url=https://search.proquest.com/docview/1987683834?accountid=14507%0Ahttp://jq6am9xs3s.search.serialssolutions.com?ctx_ver=Z39.88-2004&ctx_enc=info:ofi/enc:UTF-8&rfr_id=info:sid/ProQ%3Aabiglobal&rft_val_fmt=info:ofi/fmt:kev:

Bamiah, M. A., Brohi, S. N. and Rad, B. B. (2018) 'Big data technology in education: Advantages, implementations, and challenges', *Journal of Engineering Science and Technology*, 13(Special Issue on ICCSIT 2018), pp. 229–241.

Camilleri, M. A. (2017) 'Corporate sustainability and responsibility: Creating value for business, society and the environment', *Asian Journal of Sustainability and Social Responsibility*, 2(1), pp. 59–74. doi: 10.1186/s41180-017-0016-5.

Čolaković, A. and Hadžialić, M. (2018) 'Internet of things (IoT): A review of enabling technologies, challenges, and open research issues', *Computer Networks*. Elsevier B.V., 144, pp. 17–39. doi: 10.1016/j.comnet.2018.07.017.

Coman, C. et al. (2020) 'Online teaching and learning in higher education during the coronavirus pandemic: Students' perspective', *Sustainability*, 12(24), pp. 1–22. doi: 10.3390/su122410367.

Darling-Hammond, L. et al. (2020) 'Implications for educational practice of the science of learning and development', *Applied Developmental Science*. Routledge, 24(2), pp. 97–140. doi: 10.1080/10888691.2018.1537791.

Gil, D. et al. (2016) 'Internet of things: A review of surveys based on context aware intelligent services', *Sensors*, 16(7), pp. 1–23. doi: 10.3390/s16071069.

Gulia, P. and Chahal, A. (2020) 'Big data analytics for iot', *International Journal of Advanced Research in Engineering and Technology*, 11(6), pp. 593–603. doi: 10.34218/IJARET.11.6.2020.054.

Hassan, N. A. (2020) 'University business incubators as a tool for accelerating entrepreneurship: Theoretical perspective', *Review of Economics and Political Science*4: 33–39, ahead-of-p(ahead-of-print). doi: 10.1108/reps-10-2019-0142.

Huber, S. G. and Helm, C. (2020) 'COVID-19 and schooling: Evaluation, assessment and accountability in times of crises: Reacting quickly to explore key issues for policy, practice and research with the school barometer', *Educational Assessment, Evaluation and Accountability*. Educational Assessment, Evaluation and Accountability, 32(2), pp. 237–270. doi: 10.1007/s11092-020-09322-y.

Khare, S. and Totaro, M. (2019) 'Big data in IoT', 2019 10th International Conference on Computing, Communication and Networking Technologies, ICCCNT 2019, (March 2020). doi: 10.1109/ICCCNT45670.2019.8944495.

Krishnamurthi, R. et al. (2020) 'An overview of IoT sensor data processing, fusion, and analysis techniques', *Sensors*, 20(21), pp. 1–23. doi: 10.3390/s20216076.

Kromydas, T. (2017) 'Rethinking higher education and its relationship with social inequalities: Past knowledge, present state and future potential', *Palgrave Communications*. Springer US, 3(1), pp. 1–11. doi: 10.1057/s41599-017-0001-8.

Kumar, S., Tiwari, P. and Zymbler, M. (2019) 'IoT is a revolutionary approach for future technology enhancement: A review', *Journal of Big Data*. Springer International Publishing, 6(1): 44–48. doi: 10.1186/s40537-019-0268-2.

Mershad, K. and Wakim, P. (2018) 'A learning management system enhanced with internet of things applications', *Journal of Education and Learning*, 7(3), p. 23. doi: 10.5539/jel.v7n3p23.

Motlagh, N. H. et al. (2020) 'Internet of things and the energy sector', *Energies*, 13(2), pp. 1–27. doi: 10.3390/en13020494.

Muthuprasad, T. et al. (2021) 'Students' perception and preference for online education in India during COVID-19 pandemic', *Social Sciences & Humanities Open*. Elsevier, 3(1), p. 100101. doi: 10.1016/j.ssaho.2020.100101.

Nizetic, S. et al. (2020) 'Since January 2020 Elsevier has created a COVID-19 resource centre with free information in English and Mandarin on the novel coronavirus COVID- 19. The COVID-19 resource centre is hosted on Elsevier Connect, the company ' s public news and information', *Journal of Cleaner Production*, 2(1), pp. 3–28.

Noura, M., Atiquzzaman, M. and Gaedke, M. (2019) 'Interoperability in IoT: Taxonomies and Open Challenges', *Mobile Networks and Applications*, 24(3), pp. 796–809. doi: 10.1007/s11036-018-1089-9.

OECD. Publishing. (2013) *Education at a glance 2013: Highlights*. OECD publishing. Retrieved from http://www.oecd ilibrary.org/docserver/download/4213201e.pdf?expires=1395222439&id=id&accname=guest&checksum=F9C6674E09CE56B8C66B5B09314145B0

Oussous, A. et al. (2018) 'Big data technologies: A survey', *Journal of King Saud University - Computer and Information Sciences*. King Saud University, 30(4), pp. 431–448. doi: 10.1016/j.jksuci.2017.06.001.

Oztemel, E. and Gursev, S. (2020) 'Literature review of industry 4.0 and related technologies', *Journal of Intelligent Manufacturing*. Springer US, 31(1), pp. 127–182. doi: 10.1007/s10845-018-1433-8.

Perwej, Y. et al. (2019) 'An extended review on internet of things and its promising applications', *Communications on Applied Electronics*, 7(26), pp. 8–22. doi: 10.5120/cae2019652812.

Ramya, D. and Poongodi, O. T. (2021) 'A study on the usage of information communication technology tools in the teaching: Learning process of engineering education', 25(2), pp. 321–326.

Rapanta, C. et al. (2020) 'Online university teaching during and after the Covid-19 crisis: Refocusing teacher presence and learning activity', *Postdigital Science and Education*, 2(3), pp. 923–945. doi: 10.1007/s42438-020-00155-y.

Sarker, I. H. (2021) 'Machine learning: Algorithms, real-world applications and research directions', *SN Computer Science*. Springer Singapore, 2(3), pp. 1–21. doi: 10.1007/s42979-021-00592-x.

Sethi, P. and Sarangi, S. R. (2017) 'Internet of things: Architectures, protocols, and applications', *Journal of Electrical and Computer Engineering*, 2017(1), pp. 1–22. doi: 10.1155/2017/9324035.

Sima, V. et al. (2020) 'Influences of the industry 4.0 revolution on the human capital development and consumer behavior: A systematic review', *Sustainability*, 12(10): 61–67. doi: 10.3390/SU12104035.

Sun, A. and Chen, X. (2016) 'Online education and its effective practice: A research review', *Journal of Information Technology Education: Research*, 15(2016), pp. 157–190. doi: 10.28945/3502.

Wynn, D. C. and Clarkson, P. J. (2018) 'Process models in design and development', *Research in Engineering Design*. Springer London, 29(2), pp. 161–202. doi: 10.1007/s00163-017-0262-7.

Yang, F. and Gu, S. (2021) 'Industry 4.0, a revolution that requires technology and national strategies', *Complex & Intelligent Systems*. Springer International Publishing, 7(3), pp. 1311–1325. doi: 10.1007/s40747-020-00267-9.

CHAPTER 4

Moral, Legal, and Ethical Issues of Technology in Education

Jashandeep Singh and Simran

CONTENTS

DOI: 10.1201/9781003254942-4

4.1 INTRODUCTION

With the worldwide availability of technology today, new and difficult ethical dilemmas have arisen. Doug Johnson, an educational technology expert, acknowledges that technology can be tremendously disruptive in the classroom. He claims that our culture has set limitations for youngsters that they cannot perform because they are not emotionally ready for such activities such as drinking, driving, likewise without teaching ethical values related to technology it is unhealthy for them to engage with the internet (Mattison, 2018).

Children are fragile and have no idea how to handle real-life situations ethically by the side of technology. Cyberbullying and copyright are issues that kids are not prepared to handle. Sometimes teachers not only deal with pupils bringing gadgets to class but also with their varying levels of proficiency in utilizing them and the internet (North West Missouri State University, 2018).

Ethics in technology is a concern of technology in a practical manner. The growth of technology has given us more power to act and take actions which we have not had to before (Green, 2021).

According to the computer ethics institute, professionals teach students the following ethics regarding computer usage:

- Do not use a computer to harm others.
- Not interfering with other people's computer work.
- Not looking into other people's computer files.
- Using a computer to steal is not a good idea.
- Do not use a computer to testify falsely.
- You must not copy or use any software that you have not paid for.
- Do not use other people's computers without their permission.
- While using the computer always show respect for others.

4.2 ETHICAL CONCERNS OF TECHNOLOGY IN EDUCATION

The term 'ethics' or 'morality', a branch of philosophy, 'involves the concept of a thin line between right and wrong'. Ethics in technology refers to not breaking any norms, such as confidentiality, copyright, hacking, and many more that might harm an individual either for personal or professional benefit (Sniderman, 2020).

Ethics are the norms in any particular department according to the higher authorities that are to be followed by all people belonging to that organization. Professional ethics are principles that set the expectations for behavior for all members of an organization (Bhasin, 2020).

A broad class of websites and services that allow connecting with people of similar interest through social networking sites such as Facebook, Instagram, and many more in the acceptable use of policies, which are set by a network administrator or by school authorities for the users of such technologies for safety and security concerns to avoid any sort of bad consequences in future (Farnese, 2012).

4.3 ETHICAL ISSUES IN TECHNOLOGY

While using online websites for different purposes such as online shopping, indulging in social media, etc., companies sell personal data of customers to make unlawful deals with the third party against the norms of their own companies. As cyber use increases, knowing if your online data is protected becomes more difficult and confusing (Watters, 2021).

4.4 WHY DIGITAL ETHICS ARE IMPORTANT

It is mandatory as it depicts the inner persona of the person behind the screen with the intentions receiver is getting after receiving the message (Davis, 2021).

The following sections outline the ethical issues of most concern for individuals.

4.4.1 Confidentiality

Confidentiality permits only authorized users to access sensitive and protected data in computer systems. Specific procedures protect data from malicious invaders while ensuring confidentiality (Techopedia, 2021). It has a particular impact on clients, encouraging them to build trust because the duty of secrecy is critical in fostering positive perception in public dealing (Widmaier, 2016).

Privacy and security are interrelated in terms of ethics, which are dependent on trust. Violation of security, as well as privacy, is a threat to a person or an organization. Within breach, the norms of security also affect the trust among them and violating the ethical principles which also pushes them to disrespect the law (Lee, 2016).

Certain laws imply to protect privacy and minimize the intrusion into personal data by unauthorized individuals. Personal data refers to the personal information of any individual that is gathered by the government, by a private organization, or by any individual. The fundamental right to privacy is not explicitly stated in India's Constitution. Rather, the right to privacy has been incorporated into other legal rights, such as free speech and expressing thoughts under Article 19(1)(a) and the right to live and free will under Article 21 of the Indian Bill of Rights. However, these legal rights under the Indian Bill of Rights are subject to constraints imposed by the state under Article 19(2) of the Constitution. In addition to this, the (Indian) Information Technology Act 2000 addresses concerns, such as civil and criminal penalties for improper disclosure and misuse of personal data, as well as breaches of contractual conditions relating to personal data. A violation of confidentiality or privacy is a crime according to Section 72 of the Information Technology Act. If a person without the knowledge of the concerned individual, is accessing any online material such as books, correspondence information, documents, or other online data in pursuance of any of the powers conferred under the Information Technology Act regulations, sharing such material to any other individual is illegal and can result in up to two years imprisonment or penalty of Rs 100,000, or maybe both.

4.4.2 Hacking

Hacking is the unlawful use of online devices, such as computers, smartphones, tablets, and networks, to harm any systems, gather information on users, stealing data and documents, or disturbing data-related activities (Fortinet, 2021).

Ethical hacking is a legal activity used to check system security and detect any data breaches and network risks. To check the system, the organization's system or network permits engineers to execute actions for online security checks. This method, unlike malevolent hacking, is planned, approved, and lawful (Simplilearn, 2021).

When any computer is active on a network, a hacker can install a virus that interacts with one's computer to access personal and financial information without any agreement. In other words, a computer predator could prey on the personal information you accidentally divulge. That information includes the following:

- Stealing usernames and passwords.
- Personal financial and other details that grant access to bank cards and accounts.
- Requests for more plastic money or new account personal identification numbers (PINs).
- To make it easier to use your credit, adding themselves as an alias so they can control as a user.
- Selling your personal information to third parties for nefarious or illegal purposes (Webroot).

A person who willingly destroys or changes any computer source code or computer algorithms, harms hardware devices or network systems, or who forces someone to destroy or make changes in any computer source code, computer hardware, or network system, risks imprisonment for almost three years and a fine of two lakh rupees, according to Section 65 of the Indian IT Act. In the environment of internet technology, academics believe it is difficult to prove intent to create unlawful loss or damage. Under the IT Act of 2000, the term 'wrongful loss' isn't defined. However, Section 23 of the Indian Penal Code defines 'wrongful loss' because 'the loss of property to which the person who loses it is legally entitled by unlawful methods'. On a first conviction, the penalty for transmitting or publishing or causing to be transmitted any obscene content online is a term of imprisonment for almost two years and a fine of up to twenty-five thousand rupees (Seth, 2008).

4.4.3 Copyright

Copyright is the rights that innately pass to the creator of work, in any form like documents, audio, video, or any software. These rights have the ability to reproduce, make a copy of any work, sale copies, and publicly perform and showcase the work (Copyright

Alliance, 2021). Audio-visual work (telecasted shows, films, online videos, recorded sound, and music); written work (lectures, blogs, documents, and written music); visual works (paintings, posters, and adverts); audio-video games and computer programming software; dramatic plays and musicals. Copyright protection is applicable for a variety of works.

It is also feasible to use a copyrighted work without breaching rights:

- Credit has been given to the copyright holders.
- Refused to make money with illegal content.
- A copy of the content in question was charged for.
- 'I've seen identical stuff elsewhere on the internet'.
- 'I bought the content, which came with a print copy and an electronic copy'.
- You are recording any stuff from any of the modes such as television, cinema, or radio.
- You copied information from a book, a poster, or a photograph.
- 'No copyright breach is intended', as per the statement (Google, 2021).

The author of the work is the original holder of the copyright, according to Section 17 of the Act. In cases where an employee develops work under the contract of company or employment, the employer becomes the copyright owner.

Copyright protection from an Indian perspective is divided into two rights:

4.4.4 Economic Rights

Original theatrical, musical, cinematographic, and sound recordings as well as creative works all are under copyright privileges. In Section 14 of the Act, the owners of the aforementioned works have economical copyrights. In the case of printed, theatrical, and audio-visual works, rights are utmost to produce the work in any form, including the electronic medium, and to perform that publicly. In computer programmed algorithms, in addition to the aforementioned rights, the author has the right to sell that, place it on hire, any copy of the computer programs, in any case, either such copy has been sold previously or put that on hire. An owner has the right to reproduce an artistic work in any form, including showing a two-dimensional work in three dimensions or three-dimensional work in two dimensions, to interact with the public through the work (Sunkar, 2020).

4.4.5 Moral Rights

Section 57(1) reveals that the owner has no right to speak for any damages for any adaption of a computer algorithm that falls under Section 52(1)(aa). It needs to be emphasized that failing to depict a work or showcase it to the satisfaction of the owner is not considered an infringement of all Section rights. Expect these rights to be claimed by the owner of the work, the legal authority of the author make use of rights granted to the author of a work by Section 57(1). The right to paternity refers to the rights of an author to claim authorship of his work as well as his right to prevent others from doing so. The right of the author's integrity permits him to avoid alterations, mutilation, or other distortion of his work, also any other action related to it, that would be harmful to his reputation as well as honor (Sunkar, 2020).

4.4.6 Netiquette

'Internet etiquette' is also known as 'netiquette'. Netiquette means showing good behavior while working online, much like etiquette is a way of courteous behavior in society. Online chat, e-mail, website comments, social media, online forums, multiplayer gaming, and other modes of communicating online are all included in this category. There are no official rules or principles of netiquette, the overall concept is to treat others respectfully and politely when using the internet (Google). The following are the examples of netiquette norms to follow:

- Avoid making aggressive or disrespectful remarks on the internet.
- Do not invade the privacy of people by revealing sensitive information, images, or audio-video that they do not want to be shared over the internet.
- Never send excessive amounts of spam e-mails to others.
- Show good sportsmanship, whether you win or lose, when playing online games.
- Refrain from humiliating people online in the form of comments or by bothering or aggravating them constantly.
- When participating in online conversation or posting comments on photographs or videos, on different platforms such as YouTube or Facebook, always stay on topic.
- Do not use obscene words.
- Replying to unfavorable comments with additional abusive words comments is not a good idea. Instead, use a positive post to interrupt the loop.
- Others who assist you online should be thanked (Tech Terms, 2017).

4.4.7 Accuracy

By all means, information, especially valued online data, can be a fantastic resource. The internet is open to people who want to get information as well as those who wish to provide data. Because there is so much information available at no cost, internet users need to learn the facts of filtering when they find data online (Encyclopedia, 2019). The internet has a lot of information available worldwide, and anyone can post as per their knowledge, which does not guarantee accuracy or prevent bad data from being reposted over the internet.

4.4.8 Cyber Security Threats

This particular security threat refers to any attempt to gain access to data, disturb digital processing, or damage information. Hacktivists, corporate spies, criminal organizations, terrorist groups, lone hackers, hostile nation-states, and dissatisfied workers are possible sources of cyber dangers (University of North Dakota).

4.4.9 Types of Security Threats

4.4.9.1 Malware

Malicious software or 'malware' is designed by hackers to intentionally harm someone else's computer. Examples of this software are worms, viruses, and trojan horses (Lutkevich, 2021). Hackers can install all these on targeted devices without the owner's knowledge. Keep your software updated and back up your important files regularly to avoid any sort of loss (Norton).

4.4.9.2 Emotet

Emotet is mainly spread via spam e-mail which attract readers to open the link attached to that e-mail. These tricky mails usually send in bulk just to target the maximum number of people as their target (Expert Insights, 2020). Emotet behaves like worms that spread to other connected devices. It is one of the most costly and effective ways to harm online devices by guessing passwords. Always remain vigilant to any suspicious links and try to protect your data with strong passwords, which are hard to crack. When you find that your device has been attacked, it is advisable that you unlink from the network and clean it in isolation, then repeat this same process on any other devices connected to that same network (Malware Bytes, 2021).

4.4.9.3 Denial of Service (DoS)

A Denial-of-Service attack shuts down a machine or a network-connected device by sending the information to the target computer for a crash. The main victims of DoS are banks, commerce departments, and other organizations that handle money-related deals. A DoS attack slows down the processing of a computer or halts the entire system processing (Paloalto, 2021). Installing anti-virus software and firewalls may stop the traffic coming toward your computer (Dalmia, 2017). If your computer has been attacked in this way, you may observe that the processing has slowed down, or you may not be able to access a particular website. If you are facing such issues, contact the network authorities to determine whether it is a network or device problem. Then activate the firewall and software to protect your system from unwanted attacks (CISA, 2019).

4.4.9.4 Man in the Middle

In this, attackers sit in the center and try to interrupt the conversation by manipulating words and pretending to be a participant on either end. This break in privacy can cause damage to involved persons of the two-party communication process (Veracode, 2021). To avoid the man in the middle attack, never use public Wi-Fi connections (coffee shops, cafes, restaurants, and other open places) as well as Wi-Fi connections, without password protection. Immediately log out to all the secured websites link after use which may hold your private data (Imperva, 2021).

4.4.9.5 Phishing

Phishing attacks is digital fraud, where the perpetrator sends several e-mail to random people to trick them into giving up personal details, such as electronic money details, bank passwords, and many more. This is theft of financial-related information (Proofpoint, 2021). The mainline phishing attackers use lures, like 'Webinar from Microsoft'. To avoid phishing attacks, always check for spelling and links before entering your details and click on the link provided in the e-mail. If you receive such types of e-mails, you may want to find out the origin, whether it is phishing or safe, etc. But do not be in a hurry to share any sort of sensitive data like birth details, name, location, etc. (Josh Fruhlinger, 2020).

4.4.9.6 SQL Injection

SQL injection attacks websites or web applications that use the database created in MY SQL, Oracle, and other SQL-based software to access customers' data of customers, such as names, contact details, trade secrets, and many more (Invicti, 2021). To avoid SQL attacks, use indirect statements from users, which use parameterized queries, as dynamic queries are not as safe as stored ones. Always try to verify the entered data and use encryption and decryption keys to store sensitive data while sending and receiving it. Keep the error messages hidden from users and update the software with the newer version regularly (Rapid, 2021).

4.4.9.7 Password Attacks

Password recovery is the way of cracking the passwords by guessing the passwords using the personal details of customers by using a list of common passwords used by them (Wikipedia, 2021). Mask attacks, Dictionary attacks, Brute force attacks, Rainbow attacks are the commonly used techniques by hackers to retrieve passwords (ITPro, 2021). Biometric protection is one of the protected forms to protect the data, as no replica can be created easily. Also make sure that login passwords are strong enough so that they are note easily cracked. Always activate pen tests for your machinery; when someone attacks the machine, you can change the password before someone attacks it completely as it activates the technology of notifying devices that have been shared the data with unauthorized users (Help System, 2021).

4.5 IMPORTANCE OF CYBER SECURITY IN THE EDUCATION SYSTEM

Being a part of the technology world, we all are exposed to this new environment – especially students, as per cyber world reports most reports are from schools and colleges. So, setting an education goal at the school level is as important as other subjects in the curriculum (India Today, 2021). The education industry is one of the most affected industries by online threats.

Not only in schools, but security is also a concern at home, such as revealing the online identity to unknown people, online gaming tricks to lure children to click on links, unwanted messages or spams that attract the kids to enter the system – all are harmful to devices used by children. Parents need to keep an eye on the activities of their children with parent control activation methods, direct and daily communication with kids, teaching them to not discuss sensitive information online with random persons, and always use password-secured Wi-Fi connections with the devices that contain personal information (Boot Camp, 2021).

4.6 THINGS TO TEACH STUDENTS WHILE WORKING ONLINE

Before using technology with internet connections, students need to know a few things to protect themselves from unwanted attacks. Use strong passwords which cannot be easily hacked, keep your account private so no one, except friends, can check your posts, try to use encryption keys to secure your sensitive data, and update the security software installed in your computer (Shalini, 2019).

4.7 THE MAIN ETHICAL CONCERNS IN THE CLASSROOM USING TECHNOLOGY

Children have been pushed to the cyber world without any education; how can we expect them to behave morally during real-life situations as well as keep their identity hidden whenever required behind the screen? It is important to teach them about the dark side of not following ethical values, which may push them toward illegal activities and, even in worse situations, destroy their future. During ethics-related discussions explaining the rules and the reasons for those rules it is necessary to explain with real-life situations for easy understanding (Mattison, 2018).

It is important to teach ethics related to computer technology in a classroom before the terms related to the technology, including not to invade the privacy of a person and avoiding unauthorized access of data by breaking code to trespass the computer memory. Schools need to take these two offenses seriously to avoid the consequences as, after school, home is the place where students spend most of their time to learn. So, institutes need to put some effort into inculcating the ethics of technology (Bencivenga, 1983).

4.8 WHAT TO TEACH CHILDREN TO REDUCE AND PREVENT CYBERBULLYING

1. Never reply to any abusive messages, as the sender always expects the reaction of the victim.
2. Do not delete these messages. Keep them as evidence if needed for future legal action.
3. Ask for the help of parents and teachers who can give you a bit of expert advice.
4. Do not try to hide this situation; be open to authorities even to police if the situation is not under control (NSW Government, 2019).

4.9 WHY IT IS IMPORTANT TO TALK ABOUT ETHICS IN TECHNOLOGY

With the innovations of technology, it is a priority for an individual to learn basic ethics to use technology at its best without being involved in any antisocial activities. Hacking, to give an illustration, has both negative as well as positive sources, as using someone's computer without authorization is illegal and has strict laws against such activities, but also mixing two or more computers for experiments using programs is healthy (Hollis, 2017). It is important to teach the difference between them and not cross that thin line to avoid any source of guilt or punishment as a result (Lovegrove, 2020). Protecting themselves (children) online requires a complete discussion about security with good habits, as good habits lead to online safety and avoid hazardous threats as a result (Geonka School, 2015).

4.10 CYBERCRIME AND CYBER LAWS IN INDIA

The Information Technology Act of 2000 addresses wrongful uses of computers: activities such as theft, forgery, hacking, or attacking other computers as well as committing serious crimes, such as pornography or accessing the personal data of someone all are included in cybercrime (information security awareness).

Almost all cybercrime is punishable under the IT Act by three or less years of imprisonment. There are a few crimes, such as cyber terrorism, Section 66F, publishing material that contains sexual content, Section 67A, including child pornography, Section 67B of the IT Act, that mandates more than three years of imprisonment (Joseph and Ray, 2020).

As per the IT Act 2000, Section 65 mandates three years imprisonment for those who tamper with the source code of any computer; Section 66 mandates three years imprisonment for stealing, editing, deleting someone's system data (Hacking); Section 66A mandates up to three years plus a fine for sending offensive messages via any communication mode; Section 66B mandates almost three years plus a fine for those who are involved in intentionally stealing resources from a computer; Section 66C mandates a sentence up to three years plus fine for stealing the identity of another person; Section 66E mandates three years imprisonment with a fine for violating someone else's privacy; Section 66F mandates lifetime imprisonment for those who use a computer for terrorism; Sections 67A and 67B mandates up to 7 years for publishing both adult and child pornography (Sarmah, Sarmah, and Baruah, 2017).

4.11 HOW TO REPORT CYBER CRIME IN INDIA

To report a cybercrime, you need to go to the nearest police station where the actual crime is committed. You then need to contact the cybercrime department of your state which can be done in either online or offline mode. To do so, you need to present a server log, a copy of both original as well as compromised data with the suspect list in case of hacking. But in case of proving and abusive online message (e-mail, chat message, or any other form), it needs to be in the inbox, and the hard copy of the same needs to be submitted (Shalini, 2019).

4.12 CONCLUSION

This paper extracts that how ethics are important while working online behind the blue screen and interacting with others. However, the government sets rules and regulations that must be followed when one wants to use a computer. Defying those laws is a serious crime that might either destroy the future of an individual or end in serving jail time in severe conditions. Teachers have a great influence on their students, and they can make it easy to inculcate moral values in students. We need to educate them on how to behave being a cyber-user. We need to teach them it is really important to follow the norms of government while using internet, as it is for the benefit of the users, and disobeying the laws may force them to pay a heavy price depending upon the severity of the crime.

REFERENCES

Bencivenga J, (1983), Computer Ethics: Students Must Learn the Values, not Just Skills https://www.csmonitor.com/1983/1114/111412.html (accessed, August 2021).

Bhasin H, (2020), What Are Professional Ethics? https://www.marketing91.com/professional-ethics/ (accessed, September 2021).

Boot Camp, (2021), Cybersecurity in Education System: What Parents, Teachers, and Students Should Know https://bootcamp.berkeley.edu/blog/cybersecurity-in-education-what-teachers-parents-and-students-should-know/ (accessed, September 2021)

Copyright Alliance, (2021), What is Copyright? https://copyrightalliance.org/faqs/what-is-copyright/ (accessed, August 2021).

Dalmia P V, (2017), Data Protection Laws in India: Everything You Must Know https://www.mondaq.com/india/data-protection/655034/data-protection-laws-in-india--everything-you-must-know (accessed, August 2021).

Davis B, (2021), What is the Role of Ethics in Technology Use? https://www.mvorganizing.org/what-is-the-role-of-ethics-in-technology-use/#Why_are_digital_ethics_important (accessed, September 2021)

Ecyclopedia, (2019), Internet Data Reliability of https://www.encyclopedia.com/education/news-wires-white-papers-and-books/internet-data-reliability (accessed, September 2021)

Expert Insights, (2020), What is Emotet? https://expertinsights.com/insights/what-is-emotet-how-did-it-work-and-how-can-you-keep-your-organization-secure/ (accessed, September 2021)

Farnese R, (2012), Legal, Ethical and Social Issues in Technology https://www.slideshare.net/rachelfarnese/legal-ethical-and-social-issues-in-technology (accessed, September 2021).

Fortinet (2021), What is Hacking? https://www.fortinet.com/resources/cyberglossary/what-is-hacking (accessed, August 2021).

Fruhlinger J, (2020), What is Phishing? How this Cyber Attack Works and How to Prevent this https://www.csoonline.com/article/2117843/what-is-phishing-how-this-cyber-attack-works-and-how-to-prevent-it.html (accessed, September 2021)

Geonka School, (2015), Why School Should Give Importance to Teach Cybersecurity to its Students https://gdgoenkarohtak.com/blog/why-school-should-give-importance-to-teaching-cyber-security-to-its-students/ (accessed, September 2020)

Google, (2021), Legal Help https://support.google.com/legal/answer/3463239?hl=en (accessed, August 2021).

Green P B, (2021), What Is Technology Ethics? https://www.scu.edu/ethics/focus-areas/technology-ethics/ (accessed, September 2021).

Help Sytem, (2021), 6 Ways to Defend Yourself Against Password Attacks https://www.coresecurity.com/blog/6-ways-defend-yourself-against-password-attacks (accessed, September 2021)

Hollis K, (2017), Classrooms Technologies and Ethical Issues https://thinkspace.csu.edu.au/the independentvariable/2017/01/26/classroom-technologies-ethical-issues/comment-page-1/ (accessed, August 2021).

Imperva, (2021), Man in the Middle (MITM) Attack https://www.imperva.com/learn/application-security/man-in-the-middle-attack-mitm/ (accessed, September 2021)

India Today, (2021), Importance of Cybersecurity in the Education Sector https://www.indiatoday.in/education-today/featurephilia/story/importance-of-cybersecurity-in-the-education-sector-1770586-2021-02-18 (accessed, September 2021)

Information Security Awareness, (2021), Cyber Laws of India https://www.infosecawareness.in/cyber-laws-of-india (accessed, September 2021)

Invicti, (2021), What is SQL Injection (SQLi) and How to Prevent it https://www.acunetix.com/websitesecurity/sql-injection/ (accessed, September 2021)

ITPro, (2021), Techniques Used by Hackers to Crack the Password https://www.itpro.co.uk/security/34616/the-top-password-cracking-techniques-used-by-hackers (accessed, September 2021)

Joseph V, Ray D, (2020), Cyber Crimes under the IPC and IT Act https://www.mondaq.com/india/it-and-internet/891738/cyber-crimes-under-the-ipc-and-it-act--an-uneasy-co-existence (accessed, September 2021).

Lee W W, (2016), An Ethical Approach to Data Privacy Protection https://www.isaca.org/resources/isaca-journal/issues/2016/volume-6/an-ethical-approach-to-data-privacy-protection (accessed, September 2021).

Lovegrove M, (2020), Why We Need to Talk About Ethics in Technology https://helloworld.raspberrypi.org/articles/HW06-why-we-need-to-talk-about-ethics-in-technology (accessed, September 2021)

Lutkevich B, (2021), Malware https://searchsecurity.techtarget.com/definition/malware (accessed, September 2021).

Malware Bytes, (2021), Let's Talk Emotet Malware https://www.malwarebytes.com/emotet (accessed, September 2021)

Mattison L, (2018), Ethical Issues with Using Technology in the Classroom https://study.com/blog/ethical-issues-with-using-technology-in-the-classroom.html (accessed, August 2021)

North West Missouri State University, (2018), Ethics for Technology Use in the Classroom https://online.nwmissouri.edu/articles/education/ethics-technology-use-classroom.aspx (accessed, May 2021)

Norton, (n.d.), Malware Attacks: What do You Need to Know https://us.norton.com/internet-security-malware-malware-101-how-do-i-get-malware-complex-attacks.html (accessed, September 2021)

NSW, (2019), How to Respond Cyber Bullying? https://www.digitalcitizenship.nsw.edu.au/articles/how-to-respond-to-cyberbullying (accessed, September 2021).

Paloalto, (2021), What is Denial of Service Attacks (DoS)? https://www.paloaltonetworks.com/cyberpedia/what-is-a-denial-of-service-attack-dos (accessed, September 2021).

Proofpoint, (2021), What is Phishing? https://www.proofpoint.com/us/threat-reference/phishing (accessed, September 2021).

Rapid, (2021), SQL Injection Attacks (SQLi) https://www.rapid7.com/fundamentals/sql-injection-attacks/ (accessed, September 2021)

Sarmah A, Sarmah R, Baruah A J, (2017), A Brief Study on Cybercrime and Cyberlaws in India https://southcalcuttalawcollege.ac.in/Notice/50446IRJET-V4I6303.pdf (accessed, September 2021)

Seth K, (2008), Cybercrimes Defined under the Indian IT Act, 2000 http://www.karnikaseth.com/cybercrimes-defined-under-the-indian-it-act2000.html (accessed, August 2021).

Shalini S, (2019), How to Prevent Cybercrime in India https://www.myadvo.in/blog/cyber-crime-in-india/ (accessed, September 2021).

Simplilearn, (2021), What is Ethical Hacking? https://www.simplilearn.com/tutorials/cyber-security-tutorial/what-is-ethical-hacking (accessed, September 2021).

Sniderman B, (2020), Ethical Tech: Making Ethics a Priority in Today's Digital World https://digitalagenda.io/insight/ethical-tech-making-ethics-a-priority/ (accessed, September 2021)

Sunkar K M, (2020), Copyright Laws in India https://www.legalserviceindia.com/article/l195-Copyright-Law-in-India.html (accessed, August 2021).

Tech Term, (2017), Netiquette https://techterms.com/definition/netiquette (accessed, August 2021).

Techopedia, (2021), Confidentiality https://www.techopedia.com/definition/10254/confidentiality (accessed, July 2021)

Watters A, (2021), Ethical Issue in Technology to Watch for in 2021 https://connect.comptia.org/blog/ethical-issues-in-technology (accessed, September 2021)

Webroot, How Hackers Operate and How to Protect Yourself https://www.webroot.com/in/en/resources/tips-articles/computer-security-threats-hackers (accessed, August 2021).
Widmaier C, (2016), The Impact of Modern Technology on the Duty of Confidentiality http://www.jatl.org/blog/2016/8/29/the-impact-of-modern-technology-on-the-duty-of-confidentiality (accessed, August 2021).
The University of North Dakota, (2021), Types of Cyber Security Threats https://onlinedegrees.und.edu/blog/types-of-cyber-security-threats/ (accessed, September 2021)
Veracode, (2021), Man in the Middle (MITM) Attacks https://www.veracode.com/security/man-middle-attack (accessed, September 2021).
Wikipedia, (2021), Password Cracking https://en.wikipedia.org/wiki/Password_cracking (accessed, September 2021).

CHAPTER 5

Competencies and Values for Education 4.0

Bhavana Behal and Prateek Kalia

CONTENTS

5.1 INDUSTRY 4.0 HARBINGER OF EDUCATION 4.0

With the advent of the 21st century, society took a quantum leap in the approaches and survival maneuvers to make life more comfortable and enjoyable. The fourth industrial revolution has modernized and transformed all facets of life. Industrial revolution has evolved through four stages. Each phase of the industrial revolution came with miraculous discoveries. It has intertwined the culture and society across the globe. Lorenzo et al. (2021) have delineated the inventions and innovations ascendent in different phases of the industrial revolution, such as:

DOI: 10.1201/9781003254942-5

- The first phase of the industrial revolution: The first industrial revolution incubated coal and gas as sources of energy. Steam power in the late 18th century superseded the work based on human and animal strength. It discovered the first mechanical loom in the year 1784, leading to the establishment of the engineering domain as a discipline of study.
- The second phase of the industrial revolution: The early 20th century marks the development of mass production based on the division of labor. The invention of internal combustion engines and electricity swapped the hand-building of individual machines. The modern machine age started with the launch of assembly lines constructed with conveyor belts. Productivity and mobility became the driving force behind the swift industrialization in the world.
- The third phase of the industrial revolution: A step ahead, electricity became an indispensable part of industry and household chores. In this era, people started using information technology running on electricity peripheral to automated production. In the early seventies of the 20th century, electronic technology sprang out with a new paradigm, i.e., the cybernetic revolution. Computers emerged as the electronic device to aid humans in making decisions in all spheres of work. Transistors and vacuum tubes were left far behind the evolving miniaturization that resulted in personal desktop computers. Further, in the early eighties in 20th century the term telematics was coined (the combination of informatics and telecommunication). Here information was processed and delivered through various nodes of data networks working on a particular server. It helped to bind the world together with the string of information technology. This era was called Industry 3.0.
- The fourth phase of the industrial revolution: With the dawn of the 21st century, the world experienced the interwoven linkage across the globe with the opening of the internet in the public domain. At first it was dedicated to defense purposes only. Accumulation of an enormous pool of data on the internet facilitated artificial intelligence algorithms. This automated artificial intelligence opened a new horizon, i.e., Industry 4.0. The framework, driven by artificial intelligence, bore the most powerful tool yet – mobile computers equipped with telecommunication services. This innovation made it possible to remain connected, without physical involvement, to and from any part of the world. Industry 4.0 has configured the alliance of automatic technology and cyber technology to create the realm of the virtual world. This virtual reality in the form of association among humans, machines, and data is termed as the Internet of Things (IoT), i.e., now everything exists everywhere all the time.

The fourth industrial revolution marks the beginning of a new epoch of innovation in technology, enhancing human and machine collaboration, unveiling new business opportunities, and stimulating global economic growth. The fabrication of artificial intelligence into commercial activities has led to the smart networking of machines and data. This smart technology comprises four pivotal components: the internet of data, the internet of things, the internet of people, and the internet of services (Herold et al., 2016).

As the world has entered the fourth industrial revolution, it has been made mandatory to interact, connect, and create digital platforms and use artificial intelligence for sustainability in work and life. Advancement in the confluence of man, electronic devices, and various forms of technology has a positive edged on various sectors. The education system is one such domain that seems to be impacted greatly and stands on the verge of huge transformation in the arena of Industry 4.0. The need of the hour is to prepare sound and adequate human resources to be able to face competition on a global scale. The herculean task is to improve the quality of human resources for the economies. Technology is changing with the fast pace of revolution cascading striking impacts on the education system. The challenge before the education system is to refurbish the educational goals to produce a globally competent, open-minded, and innovative pool of individuals.

Consequently, Education 4.0 has come up as a response to the present demand of the workforce, harmonizing the skills of man and machine. Learning for the youth of today can be meaningful only when we move out of our comfort zone and try to contemplate differently and implant advanced competencies in learners. Considering this educational system as revamped in terms of holistic curriculum, digitalization of content, remote learning, and digital pedagogy to meet the need of the hour. Education stakeholders (teachers, students, administrators, and managers of educational institutions) need to acquire novel techniques of building competencies in learners so that the contemporary generation develops the accountability for sustainable living, guzzling appropriate competencies rooted in the time-tested value system.

5.2 CORE COMPETENCIES AS ESSENTIAL PREREQUISITES FOR INDUSTRY 4.0

Competencies are the state-of-art of procuring skills, knowledge, and attitudes as prerequisites for an individual to deal with challenges effectively. These sets of general competencies prepare and develop the knowledge of well-organized strategies to deal with daring and complex situations.

Thus, core competencies, grilled in the flavors of soft and hard skills, need to be inculcated in the students from the initial years of their education. Soft and hard skills develop proficiencies that individuals need to imbibe during their career. In the era of Education 4.0, basic soft skills are recognized as core competencies to be nurtured through teaching and learning activities (Miranda et al., 2021), which can be described as:

- Critical thinking, which develops problem-solving abilities by involving students in real-life practical issues.
- Cooperation through group activities, where the individual contributes his exertions, and the piling up of each group member's efforts leads to problem resolution.
- Collaboration through teamwork: here students need to weave a network for affable project collaborations.
- Communication through activities which give the scope to students to effectively express their notion through narration, visual, or in written forms on a digital

platform. Such interactive exchange of knowledge promotes the skills of intricate conversations, i.e., developing the arts of negotiation, mediation, and project discussion.

- Boosting innovation and creativity through activities that help to develop the skills of crafting, designing, and exploring aids to make them capable of creative and innovative problem elucidation.

These competencies will help student to stimulate a temperament for openness and connectivity to reap the benefit of global development (Kin & Kareem, 2019). On the other hand, hard skills, also termed 'disciplined competencies' are concomitant with precise technical know-how and task-oriented expertise. The disciplinary competencies are acquired by training and development of technology and its implementation. These competencies are comprised of three key components (Miranda et al., 2021):

- The training and development of technological knowledge fostering efficacious performance skills.
- The potential to investigate, plan, produce, and execute new technologies.
- The usage of nascent technologies for automation-based smart solutions.

Since Education 4.0 concentrates on developing a workforce for Industry 4.0 to meet current and future economic needs, these competencies will inspire them to cultivate the ability of critical and analytical thinking and acknowledging the techniques of technology for innovation.

5.3 CORE COMPETENCIES TO BE INTEGRATED IN EDUCATION 4.0

Education 4.0 proposes four prime competencies as the prerequisites for the task force of Industry 4.0:

- Technical competencies consisting of work-centric knowledge and skills for automated work culture.
- Methodological competencies comprising of skills and abilities for people management and judicious decision-making in the real-time scenario.
- Social competency, which incorporates the skills and abilities to communicate and negotiate effectively.
- Personal competencies are inborn traits, which reflect individual performance. These skills can be crafted over time with experiences and practice.

A planned curriculum is required to develop these required sets of competencies through appropriate teaching-learning patterns in educational institutions. The curriculum must infuse the competency requirements obligatory for the future world of work so that the graduates can smoothly penetrate Industry 4.0. The following characteristics are enumerated for each competency to permeate the curriculum:

5.3.1 Technical Competencies

Technical competencies expound on the acquisition of knowledge and skills, one must employ technology based tools and resources appropriately. Several recent pieces of research revolve around the impact of robotic modernization on human civilization. These published pieces of information elaborate the piercing of digital platforms and automation skills in all sorts of learning environments (Romero et al., 2015). Intensive literature review and evaluative standards report deliberately enhance technical competencies. Digital literacy is ranked highest among other skills and competencies by intellectuals. Today we need more proficiency in the use of media and information and communication technologies (ICT) than in the past few decades (Martin et al., 2016). Shute & Becker (2010) differentiate fourth industrial revolution from its predecessors because of the evolution of artificial intelligence, algorithms, and automation of the value chain. Bearing in mind such transformations, educationists have started the digitalization of course curriculum to an extensive level. As we adopt technology for education, it becomes imperative to absorb the technical savoir faire carefully to ensure impeccable allocation of the content.

In this automated ecosystem, concepts transacted to learners deem to develop their skills of analyzing, critical thinking, and synthesizing to craft a new piece of knowledge with the smithereens of relevant topics presented to them (Heinrichs, 2016; Watulak et al., 2011). Students develop the skills not only to skim the semantic content text but also to scan latent meanings underpinning the concept (Ananiadou & Claro, 2009). The interactive association of the facilitator and learner with the technology, to incorporate the participatory culture is advancing (Jenkins et al., 2009). Remote access to virtual classes on social media and mobile teaching has motivated education stakeholders to synchronize with digital proficiency. Along with classroom transition of knowledge, teachers need to upgrade their learner's technological competencies.

The technical competencies are configured on the following dimensions:

- Teaching-learning process guidelines for managing teaching aids and classrooms.
- Creating enhanced learning environments with completely digital integration (O'Flaherty & Beal, 2018).
- Various modes to communicate and collaborate with the education stakeholders (parents, students, etc.) through the digital platform (Jensen at el., 2010).
- Professional development and innovative projects surfacing on digital platforms (Tang et al., 2015).
- Concerns and issues regarding ethics, societal, economic, and human problems sprouted from the use of information and technology use.

Thus Education 4.0 needs to implant technical competencies such as interactive digital skills with participatory culture in digital forums, the art of communication in the form of e-mail, digital discussion forums, webinars, etc.; proficiency in the usage of digital tools; critical information exploration; appraisal and judicious use of open educational resources; and innovation in the virtual learning environment. Technical competencies generate the confidence to control digital gears and electronic gadgets to drive Industry 4.0 on the wheels of Education 4.0.

5.3.2 Methodological Competencies

On the ladder of competencies, methodological competencies lie on the second rung. It is defined as the implementation of specific skills and knowledge to accomplish the planned goals. Methodological competency in Education 4.0 entitles the techniques of lesson transaction which help learners to establish a connection with the learning outcomes.

In the rote learning model of education where stress is on the spewing of events and incidents, knowledge gets transferred without any innovation. In such a traditional method of teaching, knowledge flows, coupled with the teachers' interpretation of their meaning, and is installed in the minds of students without being critically mulled over. Constructive pedagogy propagated by John Dewey and his contemporaries dominated the dawn of the 21st-century education system. A decade after revolutionizing the education arena, the digital environment strengthened the methodology which encourages students to explore and research the topics directly involved. In the modern education system, teachers are playing the role of mentor and facilitator toward self-paced personalized learning (Rich et al., 2017). This has further evolved the smart digital learning system in the background of automated digital technology. The permeation of modern technology has changed the canvas of teaching methodologies. Such reforms are outlined here (Lorenzo et al., 2021):

- Substitution digital technologies have altered the use of educational tools in combination with cybernetics.
- Augmentation technology have modified the teaching strategies with the use of virtual realms.
- Modification digital tools have substituted the use of educational tools, impacting the functionality of classroom pedagogy.
- Redefinition digital has brought in changes in the dynamics of teaching and learning, giving rise to psychological and moral concerns, forming a milieu of education with modern global standards.

The fulcrum of teaching methodology rotates on information management wherein creating practical knowledge in coalition with content knowledge is the principal objective. In the continuously proliferating world of information, access to specific facts has become instantaneous on the internet. As the internet is flooded with fragments of the disrupted pieces of knowledge spread around time, space, and ideology, without drawing concerte conclusion on the topic. This leads to bewilderment.

As a result, various forms of knowledge gaps emerge. To patch these gaps, mushrooming in the digital education system, we need to evolve tools for smart learning with judicious use of artificial intelligence algorithms. Extraction of concrete knowledge, in collaboration with the subject experts and institution heads from across the globe on the digital platform, engenders progression of critical analysis, self-evaluation, and high-order thinking. The starring job of an educator in such a scenario is to weave the thread of independent debris into a systematic landscape of knowledge acquisition.

Hence methodological competency adopted in Education 4.0 should nurture the learning traits, such as self-reflection, metacognition, digital learning, self-assessment, self-motivated learning, self-reliant learner, direction toward intellectual risk, collaborative learning through open educational resources, and knowledge construction based on self-collected and authenticated shreds of evidence.

5.3.3 Social Competencies

Snowballing information, speedy vicissitudes of technology, and globalization determine and establish the educational strategies. Consensus is growing around the world that technology is an architect of new work culture and the prospects that have emerged. Industry 4.0 has restructured the demographic profile impacting work vicinities. It is being anticipated that there is increasing intricacy in work profiles, along with a growing necessity for a global job environment. Therefore Education 4.0 is needed to develop a prolific ecosystem with protruding social competencies. Social competence confers the social, emotional, and cognitive skills and behaviors individuals adapt for better adjustment in society. Social competencies can be accorded with four skills, i.e., communication skills, cultural awareness, global awareness, and leadership.

Education 4.0 should incorporate and promote such abilities; recent studies can be traced for the exploration of in-depth traits of social competencies. Researchers have thoroughly discussed communication skills as the main ingredient for collaboration in the global environment of Industry 4.0 along with other prominent skills (Chalkiadaki, 2018). Emphasis is laid on the oratory and scripting skills, effectively, to lure the global partners (Teruggi & Zuccoli, 2015). Besides giving priority to mother tongue, multilingual communication is becoming the reality of Education 4.0. Hence the need arises to foster team-working skills to make a heterogeneous environment more congenial for effective communication. In addition, conflict management skills have also become an indispensable acquisition (Ananiadou & Claro, 2009). In the last decade, researches on social competence have proved that socially competent individuals have better psychological well-being and achievements. On the contrary, socially incompetent individuals meet failures in all walks of life. Dissatisfaction and distress with social relationships are prognosticators of maladjustment, including an unfriendly attitude toward society.

Hence the objective of Education 4.0 revolves around developing skills which help to establish positive relationships, such as communication and collaboration in local, national, and foreign languages through written and verbal modes; working with heterogeneous teams; having an open mind to management people and conflicts; nurturing cultural awareness to appreciate diverse beliefs and build inter-culture relationships; and taking initiative, self-motivation, and the will for independent entrepreneurship.

5.3.4 Personal Competencies

In the mutable nature of the world, where every minute globalization is transmuting the socioeconomic scenario, we need an education system that raises a class of individuals who can safeguard the inherited landscape with sustainable modernization. Therefore, personal competencies should be drilled in, cautiously enhancing their intelligence, social, and emotional quotients.

Personal competencies are examined by different scholars in the sphere of commercial and educational activities. The creative way of thinking is one of the vital personal skills in the context of Education 4.0. Academic objectives must yield creative outputs (Martin et al., 2016). Curiosity and imagination add to playfulness, which keeps the energy high for productive work (Teruggi & Zuccoli, 2015). It promotes the ability to innovate 'cocreativity', as given by Romero et al. (2015), which describes the dimension of collective creativity to produce collaborative work.

Significant abilities to be assimilated with personal competencies, explored in the intensive literature review by Chalkiadaki (2018) are self-direction and self-reflection as essential skills for the students of Education 4.0. Further, the protection of rights, beliefs, values, interests, and needs with the intention of developing and steering personal and professional life, is the key to survival in the era of the Fourth Revolution (Ananiadou & Claro, 2009). To enumerate more characteristics of personal competencies – adaptability, agility, integrating intellectual risk-taking, and people management skills (Abdullah & Osman, 2010) are accentuated to it. Innovation with direct solutions is also valued for effective outcomes (Cruz & Orange, 2016) of the work. Scrutinizing retrieved facts and data leads to critical synthesis and evaluation. Higher-order thinking is measured as an indispensable skill for the individual and personnel of the 21st century along with rigorous rational reasoning, analysis, and the talent to make cognizant decisions (Abdullah & Osman, 2010). To chisel the required characteristics of the individual for Industry 4.0, these skills are to be nurtured in them.

Another fundamental skill of personal competencies can be perceived as problem-solving. Grooming youth with the essentials of problem-solving at their tender age prepares them to successfully overcome the impediments of social and professional life. Cultivating virtuous problem-solving skills develops the traits of independence, confidence, and social acceptability. An efficient problem solver can effectively cope with emotional and mental mayhem. Problem-solving is also often associated with the ability of analytical thinking.

Consequently, Education 4.0 should aim to carve the personal competencies, such as adaptability, continuous improvement, ability to interact with modern interfaces, holistic analytical thinking, self-motivation, self-reliance, self-awareness, self-management, investment in self-development, confidence to take charge of a life plan, and emotional intelligence.

5.4 VALUES TO BE EMBEDDED WITH CORE COMPETENCIES

Values are the belief system and ideologies of people, the basis on which they make decisions in all aspects of their life. Values are the guiding principles for the course of action people take in their personal and professional arena. Competencies can only yield progressive outcomes when the modern education system is nurtured with values embedded in it.

Various international forums have propagated the importance of inculcating attitudes and values in the school curriculum. Values articulated in these international forum instruments (such as Sustainable Development Goal 4.7 on Education (Ferguson & Roofe, 2020; UNESCO-UIS, 2018; United Nations Educational, Scientific and Cultural Organization (UNESCO), 2017; United Nations, 2015), G7 Summit Leaders' Declaration 2016, OECD Global Competency Framework (24OECD, 2019; OECD, 2018; OECD, 2016), the UN Charter, the Council of Europe's reference Framework of Competencies for Democratic Culture (Barrett, 2020), and the Millennium Declaration) are committed to foster multiculturalism for strengthening global connectivity and to restore faith among communities. Even though the terms used to communicate the values in the above-mentioned documents are not alike, a deductive conclusion can be made on the agreed-upon values to be groomed in learners, such as justice, responsibility, tolerance, cultural diversity, equality, freedom, global-mindedness, and democracy. Thus, the baton is placed in the hands of educational institutions to rekindle the lamp of values and beliefs for global sustainable living. Education 4.0 owns the onus for sturdier efforts to nurture a value system comprised of global citizenship at the micro-level to carve intensively rational societies at the macro level.

The focal aim of education is to make students responsible, accountable, and productive citizens of a country. In the last decade, the canvas of society has changed its color drastically. Now the focus has drifted toward global citizenship. Hence the liability is on the school and educators to prepare contributing members for the global community. They need to germinate a value system for global sustenance. As we know, the seeds of values are sown in the initial years of elementary schooling, which sprouts gradually. Schools should provide a healthy environment for nurturing values through various teaching-learning activities. The values enshrined in the above-mentioned international instruments can be floated as a hidden curriculum in the classroom teaching and through extracurricular activities, such as are outline in the following sections.

5.4.1 Human Dignity and Respect

The vision of the school prospectus should reflect human dignity as a core value and stakeholders must adhere to it. Teachers should acknowledge everyone as a unique entity and take care of students without any discrimination. The principles of human dignity are grounded on tenets that, irrespective of any differences in terms of gender, class, creed, or belief, all people are tied together with the same value of humanity. Thus, our religious and regional diversity should not hurdle respect for our fellow beings. The value of human dignity ingrained in educational curriculums should instigate the establishment and generate respect for human rights. We are born with certain rights as humans. Schools should provide the opportunity for the learners to unfold the history of these rights. Creative writing, role play, morning assembly talks, and contemplation on such issues will certainly shore up the respect for incongruent human existence.

5.4.2 Equality, Tolerance, and Freedom

The awareness and acknowledgment toward the experience of our differences must be molded toward appreciation of diversity and entwining the feeling of tolerance. Equal opportunities must be given to all, so they can take self-decision of their choices and freely express their opinions. Schools must create an ample scope of accommodation for diversity and ascertain the sense of belongingness among the students. Schools should arrange community lunches, visit social organizations, spending a day in a nearby rural area, interviewing about the living standard of domestic servants or some poor person in their proximity, and debating and discussing such topics. It will help to ignite the value of tolerance for others and help to comprehend the diverse social status of the people and give them the insight to infer that miseries evolve from inequality. Further, encourage them to make provisions for the eradication of inequalities fostering integrity for healthy relationships in the society. Tolerance advances social relationships as valuing others' existence, and opinion are indispensable for constructive relationships. In the environment of equality and tolerance, germination of self-respect and confidence to aim for higher goals takes place. As studies reveal, the students who possess high self-respect have better academic outcomes. Self-respect also develops the conscious level to accept the situation and forgive themselves and others (Dillon, 2001).

5.4.3 Justice

Justice is thoroughly interwoven with the yarn of equality. It is essential for the individual and social well-being of society. Valuing justice promotes tolerance and helps to erase stereotyping and prejudice. To indoctrinate the sense of justice among budding global citizens, traits of critical thinking, problem-solving, self-evaluation, decision-making with a rational outlook should be promoted. Teachers should be competent enough to detect student's mental or physical conflicting situations and promptly provide counseling. Here it becomes imperative to nip the cause of conflict, as with caution prevails the sense of justice. Adolescents inheriting belief systems, deep-seated with the values of justice and equality exhibit high social and emotional quotients; as a result, they accomplish more in academic and school performance. Additionally, the value of justice is required for endorsing multiculturalism across the globe (Lerner, 2015).

5.4.4 Global-Mindedness, Responsibility, and Cultural Diversity

Values of justice and equality cultivate the feeling of belongingness toward community, nation, and the global society. This sense of connectivity is echoed through an individual's attitudes, beliefs, and actions. It helps to inculcate the sentiment of being accountable and responsible for the world community which progresses the value of global-mindedness. An extremely global-minded person respects and accommodates diversity and differences experienced in society. It incites them to ascertain and address the economic and social disorders across the world. Accordingly, a cross-cultural environment with the philosophy of global-mindedness is based on the following parameters (Demello, 2011):

- Responsibility: being accountable for the social constructive welfare of the world community along with economic development.
- Cultural pluralism: the existence of multiple cultures to be comprehended as each custom and faith adds some values to humankind. Core values from each belief system to be appreciated for the networking of global society.
- Efficacy: the presence and involvement of every single individual in national and international issues can make a huge difference in total output.
- Global centrism: setting goals for the long-term considering global welfare while making progress at the local and national fronts.
- Interconnectedness: Fabricating socially and economically interlinked culture, so that a sense of global human family is developed.

The objective of global-mindedness clinches the notion of promoting respect for fellow humans and viewpoints of different cultures (Hill, 2012). Multicultural education and cultural diversity promote social inclusion at the international level. To promote this value, schools should send students for literary activities and sports events at international levels. Plan international tours, participate in intercultural programs, and make provisions for cohabiting of students from different regions in one campus for an event. Cultural pluralism establishes the testimony of understanding between the diverse cultural backgrounds which exists in our society. However, for intercultural understanding, one should not substitute the sense of nationhood with something high on moral grounds (Heyward, 2002).

5.4.5 Democratic Values

Democratic values are the ideologies that consider society fair and equal for all. Here an individual can take the initiative, freely express himself, seek justice, participate in the decision-making process, and defy unproductive policies and norms. These values should be cultured in students via the school environment. Teacher's behavior and gestures should reflect democratic attitudes, classroom activities need to be scheduled with the suggestion of students, class responsibilities among peers should be equally distributed according to their skills, and cooperative and teamwork projects should be encouraged. School curriculum, teaching-learning activities, and other resources used in the school premises shall deploy the tents of democratic principles. School leadership should work as the savior of democratic values. Democratic values thrive to nourish in students' qualities like, leadership, respect for diversity, tolerance for distinct opinions, collaboration for teamwork, sensibility, and, value for life. In schools, teachers should use academic content to skills of competencies and value system at every level and in each class. Industry 4.0 has set the clock for education personnel to train tech-savvy nation builders.

5.5 CONCLUSION

Endeavors for development mandate the functionality of all occupational sectors conscientiously. Thus, the onus is on the education system to feed the needs of the task force required by Industry 4.0. Schools, colleges, and universities should pay heed to the inculcation of core competencies and values essential for sustainable global peace and progress in all spheres of life. Since the education system has been transformed from a rote learning scenario to an automated platform. Humans have become the architect as well as a pupil of the machine. The world is witnessing not only a change in pedagogical strategies but also a huge remodeling in the perspective of education as a concept. This urges to revamp the education structure from the grassroots level. Infusion of policies that promotes the practical application of knowledge, teachers' orientation beyond textbook content, real-life problem handling tactics, self-assessment ability, physical and mental wellness should be indispensable for the educational institution at all levels. Further, the curriculum components should be blended with the principles of global-mindedness, democratic stability, human dignity, tolerance, equality, freedom, and justice for members from all strata of the society. Simultaneously the seasoning of technological competencies should be channeled through methodological competencies for maturing social and personal competencies in the learners for healthy productive outcomes. This will help an individual to grow the ability for critical thinking, positive social interaction, and collaborative work in the environment of fourth industrial revolution.

REFERENCES

Abdullah, M., & Osman, K. (2010). 21st century inventive thinking skills among primary students in Malaysia and Brunei. *Procedia: Social and Behavioral Sciences*, *9*, 1646–1651. https://doi.org/10.1016/j.sbspro.2010.12.380

Ananiadou, K., & Claro, M. (2009). 21st century skills and competences for new millennium learners in OECD countries. In OECD Education Working Papers (No. 41; Issue 41). https://doi.org/10.1787/218525261154

Barrett, M. (2020). The Council of Europe's reference framework of competences for democratic culture: Policy context, content and impact. *London Review of Education*, *18*(1), 1–17. https://doi.org/10.18546/LRE.18.1.01

Chalkiadaki, A. (2018). A systematic literature review of 21st century skills and competencies in primary education. *International Journal of Instruction*, *11*(3), 1–16. https://doi.org/10.12973/iji.2018.1131a

Cruz, M., & Orange, E. (2016). 21st century skills in the teaching of foreign languages at primary and secondary schools. *The Turkish Online Journal of Educational Technology*, Special issue, 1–12.

Demello, M. A. (2011). *The Impact of Study Tours in Developing Global-Mindedness among PK-12 Educators in Sotheastern Massachusetts*. Boston, MA: Northeastern University.

Dillon, R. S. (2001). Self-forgiveness and self-respect. *Ethics*, *112*(1), 53–83. https://doi.org/10.1086/339140

Ferguson, T., & Roofe, C. G. (2020). SDG 4 in higher education: Challenges and opportunities. *International Journal of Sustainability in Higher Education*, *21*(5), 959–975. https://doi.org/10.1108/IJSHE-12-2019-0353

Fleming, E. C., Robert, J., Sparrow, J., Wee, J., Dudas, P., & Slattery, M. J. (2021). A digital fluency framework to support 21st-century skills. *Change: The Magazine of Higher Learning*, 53(2), 41–48, DOI: 10.1080/00091383.2021.1883977

Heinrichs, C. (2016). Exploring the influence of 21st century skills in a dual language program: A case study. *International Journal of Teacher Leadership, 7*(1), 37–57.

Herold, G., Practice, G., & Industrial, L. (2016). *Leadership in the Fourth Industrial Revolution.* Stanton Chase. Retrieved 11 September 2021 from https://executiveacademy.at/fileadmin_synced_%0Aassets/documents/White_Papers/Leadership-in-Fourth-Industrial-Revolution-Stanton%02Chase.pdf

Heyward, M. (2002). From international to intercultural. *Journal of Research in International Education, 1*(1), 9–32.

Hill, I. (2012). Evolution of education for international mindedness. *Journal of Research in International Education, 11*(3), 245–261. https://doi.org/10.1177/1475240912461990

Jenkins, H., Purushotma, R., Weigel, M., Clinton, K., & Robison, A. J. (2009). *Confronting the Challenges of Participatory Culture: Media Education for the 21st Century.* Cambridge, MA: MIT Press.

Jensen, P., Paige, C., Sweredoski, D., & Yanoff, E. (2010). Reading and responding in the 21st century: Digital literacies and the NYSRA Charlotte award. *Language and Literacy Spectrum, 20*, 15–23.

Kin, T. M., & Kareem, O. A. (2019). School leaders' competencies that make a difference in the era of education 4.0: A conceptual framework. *International Journal of Academic Research in Business and Social Sciences, 9*(4), 214–225. https://doi.org/10.6007/IJARBSS/v9-i4/5836

Lerner, R. M. (2015). Promoting positive human development and social justice: Integrating theory, research and application in contemporary developmental science. *International Journal of Psychology, 50*(3), 165–173. https://doi.org/10.1002/ijop.12162

Lorenzo, N., Gallon, R., Palau, R., & Mogas, J. (2021). New objectives for smart classrooms from industry 4.0. *Technology, Knowledge and Learning, 26*(4), 719–731. https://doi.org/10.1007/s10758-021-09527-0

Martin, C. K., Nacu, D., & Pinkard, N. (2016). Revealing opportunities for 21st century learning: An approach to interpreting user trace log data. *Journal of Learning Analytics, 3*(2), 37–87. https://doi.org/10.18608/jla.2016.32.4

Miranda, J., Navarrete, C., Noguez, J., Molina-Espinosa, J.-M., Ramírez-Montoya, M.-S., Navarro-Tuch, S. A., Bustamante-Bello, M.-R., Rosas-Fernández, J.-B., & Molina, A. (2021). The core components of education 4.0 in higher education: Three case studies in engineering education. *Computers & Electrical Engineering, 93*(September 2020), 107278. https://doi.org/10.1016/j.compeleceng.2021.107278

OECD (2016). PISA 2018 draft analytical frameworks, May 2016. Retrieved 12 September 2021 from https:// www.oecd.org/pisa/data/PISA-2018-draft-frameworks.pdf

OECD (2018). PISA, preparing our youth for an inclusive and sustainable world: The OECD PISA global competence framework. Retrieved 12 September 2021 from https:// www. oecd. org/ educa tion/ Global- compe tency- for- an- inclu sive- world. pdf

OECD (2019). *Attitudes and Values for 2030.* OECD Future of Education and Skills 2030. Retrieved 12 September 2021 from www.oecd.org/education/2030-project

O'Flaherty, J., & Beal, E. M. (2018). Core competencies and high leverage practices of the beginning teacher: A synthesis of the literature. *Journal of Education for Teaching, 44*(4), 461–478. https://doi.org/10.1080/02607476.2018.1450826

Rich, P. J., Jones, B., Belikov, O., Yoshikawa, E., & Perkins, M. (2017). Computing and engineering in elementary school: The effect of year-long training on elementary teacher self-efficacy and beliefs about teaching computing and engineering. *International Journal of Computer Science Education in Schools, 1*(1), 1–20. https://doi.org/10.21585/ijcses.v1i1.6

Romero, M., Usart, M., & Ott, M. (2015). Can serious games contribute to developing and sustaining 21st century skills? *Games and Culture, 10*(2), 148–177. https://doi.org/10.1177/1555412014548919

Shute, J. V., & Becker, J. B. (2010). Innovative assessment for the 21st century. In V. J. Shute & B. J. Becker (Eds.), *Springer.* New York: Springer. https://doi.org/10.1007/978-1-4419-6530-1

Tang, S. Y. F., Wong, A. K. Y., & Cheng, M. M. H. (2015). The preparation of highly motivated and professionally competent teachers in initial teacher education. *Journal of Education for Teaching, 41*(2), 128–144. https://doi.org/10.1080/02607476.2015.1010875

Teruggi, L. A., & Zuccoli, F. (2015). The status of twenty-first century skills within the University of Milan-Bicocca's Degree Programme in Primary Education. *E-Pedagogium*, 15(2), 75–87. https://doi.org/10.5507/epd.2015.018

UNESCO. (2017). UNESCO moving forward the 2030 Agenda for Sustainable Development. UNESCO, Retrieved 22 September 2021 from https://en.unesco.org/creativity/sites/creativity/files/247785en.pdf

UNESCO-UIS. (2018). Quick guide to education indicators for SDG 4. UNESCO Institute for Statistics. Retrieved 12 September 2021 from http://uis.unesco.org/sites/default/files/documents/quick-guide-education-indicators-sdg4-2018-en.pdf

United Nations (2015). Transforming our world: The 2030 agenda for sustainable development, Retrieved 12 September 2021 from https:// susta inabl edeve lopme nt. un. org/ conte nt/ docum ents/ 21252 030% 20Age nda% 20for% 20Sus taina ble% 20Dev elopm ent% 20web. pdf

United Nations Educational, Scientific and Cultural Organization (UNESCO) (2017). Education for sustainable development goals learning objectives, Paris: UNESCO, Retrieved 12 September 2021 from www.iau- hesd.net/sites/default/files/documents/247444e.pdf

Watulak, S. L., Laster, B. P., & Liu, X. (2011). Technology stalled: Exploring the new digital divide in one urban school. *Journal of Language and Literacy Education*, *7*(2), 1–21.

CHAPTER 6

ICT Tools for Language, Literacy, and Early Childhood

Aastha Sinha, Mahima Bhardwaj,
Parul Johri and Sachidanand Singh

CONTENTS

DOI: 10.1201/9781003254942-6

6.1 INFORMATION AND COMMUNICATION TECHNOLOGY IN EDUCATION

The terminology Information and Communication Technology (ICT) alludes to telecommunications-based information access technologies. It's indistinguishable from information technology (IT), but it's particularly concerned with communication technologies. This can be seen in online platforms, wireless connections, communication devices, and other avenues of interaction. According to the United Nations Educational, Scientific, and Cultural Organization (UNESCO), Information and Communication Technology (ICT) is a science-based, analytical, and technological field as well as a management strategy for managing information, its use, and moral implications, and socioeconomic and contextual factors. It means that we now have plenty of opportunities to employ ICT in the education sector, specifically teacher training programs, and increasing teacher quality in order to effectively teach technology (Pearson 2003). The educator is the most significant part of education in today's society. They continue to work in all areas for the benefit of our society. Students can be sculpted into good social workers, politicians, poets, philosophers, and other members of society by skilled teachers. Skilled educators have the ability to mold students. Teachers have the ability to communicate with kids in a pleasant manner. Technology's rapid evolution has resulted in innovative changes in how we live and societal expectations.

In the academic cycle of the 21st century, Information and Communication Technology (ICT) in education has flourished in popularity beyond mere mention; its adequacy has earned a variety of accolades and has made it one of the 21st century icons that allure academicians and researchers' attention (Clark et al. 2016). Acknowledging the influence of digital technologies on industry and routine activities, educational institutions are revamping their curriculums and student development to bridge the technological gap between today's teaching and learning and tomorrow's teaching and learning.

ICT includes competencies, programming, technologies, and infrastructure. The four components of ICT education are information and communication technological proficiency, ICT support and funding of advanced technologists, advanced industry and government uses of ICT, and ICT use in science and technology by researchers. To ensure that ICT is widely used, expert technologists should be engaged to support ICT/digital literacy by training teachers and developing learning methods for students (Molnar 2007). The impact of ICT is becoming increasingly noticeable in classroom teaching.

Society is driving educational institutions to respond successfully to technology changes since ICTs give both learners and lecturers additional possibilities for personalizing learning and instruction to individual requirements. ICT education is represented with different categories and classifications which need to be learned before entering the underlying depth of ICT-based education (Hennessy et al. 2005):

1. **Blended Learning** is an expression used to designate learning paradigms that integrate face-to-face educational experience with digital training. A teacher, for example, can encourage student learning in the classroom while also utilizing the paradigm (modular object-oriented dynamic learning environment, MOODLE) in order to enable learning outside the classroom.

2. **E-learning** is an element of a training program that uses an information network, such as the internet, a wide area network (WAN), and a local area network (LAN) for curriculum presentation, interaction, and/or furtherance. Web-based learning is a sort of online education that entails utilizing an internet browser to imitate, blackboard, or internet explorer to learn (Gimenez 2000).

3. **A learner-centric** education is the constructivism training paradigm that inspires an environment in which the emphasis is placed on the knowledge, abilities, perspectives, and convictions that trainees are responsible for bringing to the training techniques. In the context of this article, it refers to a student's personal involvement in a learning endeavor while using a laptop and/or a web-based connection.

6.1.1 Information Communication Technologies in Edification Forum

ICT assists and challenges educators in the newest innovations by incorporating various technologies. The World Wide Web (www), is the supreme and most frequently used internet service (along with IRC and e-mail). Its renown has skyrocketed because it is so simple to work with and rich with information (Chun 2003).

Vijiyarani and Suganya (2015) reported the application of the web as a collection of interrelated documents saved on the internet, Websites, or computer sites (Vijiyarani and Suganya 2015). E-learning, also known as 'online training'. E-learning can be used to educate at all levels, including formal and informal learning. The cyber infrastructures, E-portfolios, and other components are also included. All of the components listed below work together to create a user's digital identity and connect all educational stakeholders. It also makes interdisciplinary experimentation easier.

- **Group Conversation** – internet relay chat (IRC) is one of the crucial online services for live talking. Through the internet, a circle of people with a familiar interest may rapidly swap their opinions and vision. The internet technologies that are required to make ICTs easier to use for education are outlined (www, video conferences, teleconferences, word processor, mobile conferences, CD database, Intranet, internet, etc.).
- **E-Modules** – using a word processor that the user can access over the web, written components are translated into cloud storage format in the computer network.

To fully harness the potential of current information and communication technologies (ICTs) for educational purposes, the basic requirements that are essential to meet are described as follows:

- In their lectures, educators, institutions, school administrations, and learners must have adequate access to digital tools and the online databases (Collis and Jung 2004).
- Teachers and students need access to premium quality, significant, and culturally appropriate web-based context.
- To help all children reach high academic standards, teachers must have the knowledge and abilities to use new digital tools and resources.

- Teachers of the future will be able to efficiently use new learning tools in their classrooms. This difficult job necessitates the acquisition of new assets, knowledge, and meticulous preparation for many teachers' educational programs (Loveless 2011).

When approaching this task, it's important to consider the following: technology's impact on global society and its educational consequences; the process of learning for the learners which includes a large body of information; the designing of more effective and entertaining learner-centric education environments (Yunus 2007).

The phases of educator's enhancement, and the extent to which they have adopted ICTs, context, leadership culture and foresight, learning for the entire life, and the transition procedure must all be considered when integrating technology into teacher education (Figure 6.1). Teachers are expected to have ICT capabilities in curriculum, pedagogy, technical issues, social concerns, cooperation, and networking. Establishing standards in order to supervise the educators about the role and usefulness of ICTs in faculty enlightenment is a kind of compulsory task.

There are requirements for integrating ICTs into teacher professional development successfully, as well as critical approaches to consider when integrating ICTs into education systems and managing the knowledge transfer (Hashim 2010). De Kadt et al. reported in their manuscript about the groundwork for incorporating ICTs into the curriculum development and defines the standards for technology amalgamation to be successful. It also includes various case studies that demonstrate different strategies for integrating ICTs into teacher education, as well as advice on how to create an excellent strategic digital transformation strategy. Finally, it emphasizes the importance of organizing and developing the change process, as well as getting widespread support from all stakeholders, in order to reach the aimed targets of introducing ICTs into the education planning programs for students (De Kadt et al. 2001).

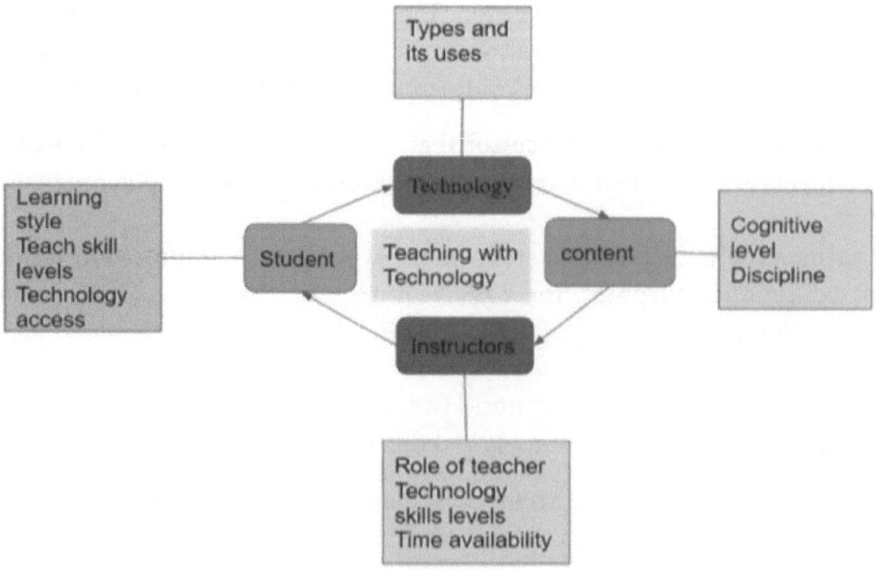

FIGURE 6.1 Teaching with technology (from Zhu & Kaplan (2001) McKeachie's Teaching Tips).

6.1.2 Educational Technology as a Concept

The use of both physical hardware and pedagogical theory is referred to as educational technology. It covers a wide range of topics, including learning theory, computer-based training, online learning, and mobile learning (m-learning) (Tinio 2002).

As a result, there are numerous distinct components to explaining educational technology's intellectual and technical development:

- The philosophy and practice of educational methods to learning are referred to as educational technology.
- Educational technology is defined as technical instruments and media that aid in the communication, development, and sharing of knowledge.
- Educational technology is used in Learning Management Systems (LMS), in particular tools for learners and educational program management, and education management information systems (EMIS). Such courses may be referred to as 'Computer Studies' or 'Information and Communication Technology'.

6.2 THE APPLICATION OF INFORMATION AND COMMUNICATION TECHNOLOGIES (ICT) IN TECHNICAL EDUCATION

In technical education, ICTs are essential instruments. Electronic management (e-management) helps organizations make better decisions (Zain et al. 2004). For the need for ICT in technical education 'the strength to plan various activities and to conduct these activities should be efficient so that the planning could be applied effectively', as per the International Institute for Communication and Development (IICD) (De Kadt et al. 2001).

The majority of the data is in textual format and therefore difficult to access (Picciano 2002). Teachers' earnings, student results, the strength of each class, and the data collected by analyzing the statistics in general are all scattered and difficult to come by. School administrators must have the necessary knowledge, skills, and awareness of the pedagogical and social aspects of ICT amalgamation. The educational dimension relates to how ICTs are used in everyday social interactions, whereas the social dimension refers to how ICTs are employed in curricular, professional, administrative, and economical aspects (Neo and Neo 2001).

It mainly supports administrators in acquiring a clearer understanding of the scale of the educational system. ICT should be actively used in educational administrative offices (Cohen 1976). Studies on the use of ICT in Technical Education have revealed significant accomplishments: for example, According to a study on future school heads' use of e-learning software in academic management and leadership, educational headship (e-headship), a collaborative process that brings teachers, students, and parents together to improve education quality, is always impactful in boosting teaching and learning methodologies to a greater extent (Floris 2014).

ICTs assist administrators in effectively performing school duties (Tinio 2002), increasing and providing easily accessed information to administrators (Stepp-Greany 2002), and creating a very favorable work environment. As a result, according to UNESCO IITE (2011)

'ICT facilitates technical education, the provision of learning context, and sharing of information between educators and learners'. In education, information technology helped to develop and preserve knowledge (Stepp-Greany 2002). It allows learners to access material in their own environment, avoiding the need for physical presence in order to obtain knowledge, and it puts universal knowledge in the hands of the student.

6.2.1 Information Gathering

The oldest technique of instruction has been supplanted by information technology. With a single mouse click, learners can access worldwide information properly and instantly, thanks to modern internet technologies. Audio and video systems make it easier to approach specialists and obtain information. Information technology helps to collect and exchange information from professionals all around the world through interactive learning in the form of enhanced video conferencing (Green 2007; Chenoweth 2006).

6.2.2 Encouraging Students to Learn

Knowledge technology promotes learning by allowing access to related information from any location at any time, removing the need for a physical location, and shortening class time. Self-motivated students can accelerate their learning through self-study, and they also have the potential to promote learning by bringing in the greatest teachers to the classroom (Borthwick and Gallagher-Brett 2014).

6.2.3 Assisting Educators

Education is a part of learning, and excellent teaching is the foundation of successful learning. Information technology has aided instructors in updating and improving their expertise through the exchange of ideas, acquiring information from experts, and another means (Chun et al. 2016).

6.3 USE OF ICT TO ASSIST LANGUAGE EDUCATION

ICT has grown more important in language acquisition (Markee 1997). Its application in school has aided in the development of linguistic skills. I believe that ICT should be included into the curriculum to aid students and teachers in the teaching and learning of languages

Language teachers throughout the world have recently been discussing ICT. It is a proportionately new circumstance to employ ICT in language training. Furthermore, educational academics have been interested in it for over two decades in order to figure out a way to overcome the barriers to adopting ICT and to incorporate it into the curriculum (Dhital 2018) .

According to surveys, language experts are focusing on implementing the use of ICT in language acquisition (Pun 2013). Language education using advanced technologies is used to help students achieve their goals and have a better understanding of the target language. This covers things like using the online services, downloading YouTube program videos, video recording display, iPad apps, and learning activities like e-learning, according to the use of digital and interactive technology (Altun 2015).

In this regard, the study found that learning with multimedia is more practical and appropriate for usage in any subject, as well as being very simple to employ in a learning session. In order to increase students' interest and hence develop enthusiasm in the learning process, creative educators will frequently seek out new teaching ideas and devise fascinating teaching aids (Laudari 2019).

Language teaching and learning using digital technologies is becoming more common among students; teachers must equip themselves with the most up-to-date technology in order to make their lessons meaningful (Chun et al. 2016; Kessler 2018). Smart boards may be quite useful in the classroom for engaging and encouraging pupils. For example, even if the teacher is not an indigenous speaker, learners can effortlessly learn how to pronounce words utilizing a smartphone/board (Yunus et al. 2009).

Teachers can use modern pedagogy with the help of technology, which helps the language learning process. Technology is becoming increasingly significant as a strategy to help teachers encourage their pupils' linguistic skills (Wishart et al. 1999; Atas et al. 2019).

To improve their teaching, learning activities, and relevance, today's language teachers must understand how to use technology to their advantage and integrate it into their teaching methods. According to Dudeney 2000, the incorporation of advanced computer software has aided students in strengthening their language grasping power (Dudeney 2000).

Gage 2013 discussed the efficacious usage of forefront techniques for increasing language learners' abilities. The internet's invention, as well as collaborative software, have opened up new possibilities for learners in terms of context delivery, the ability to efficiently talk with children and with the rest of the world, the design and delivery of teaching activities, as well as the administration of the preparation of entire course and the evaluation of students (Gage 2013).

according to Al-Jarf 2004, Technology, improves writing and gives language an additional power not just by broadening the spectrum of human interpretation, but also by providing a means for documenting and accumulating knowledge. Writing is the foundation for most of today's web-based expression of thoughts, and it's crucial to note that it's just as much a part of technology as it is of sign systems (Al-Jarf 2004).

What is written and read by us is different: in each case, each person brings their own thoughts and material properties, emotions, procedures, affordances and restrictions, and so defines a unique relationship between writers and texts.

Incorporating a variety of ICT tools and applications into teaching and learning is always a good strategy. Some of these tools and applications are intended for educational use only, while others are intended for general use. Various learning philosophies can be linked to resource selections and how they are used to interpret and communicate learning benefits from the use of ICT. The use of two key approaches, namely directed and constructive instructing methodologies, has raised the usage of ICT in education (Mahnke 1996).

6.4 TOP ICT TOOLS TO DEVELOP LITERACY

In education, electronic devices are utilized to support curricula (Figures 6.2 and 6.3). The following sections describe the most often utilized communication technology for educational purposes.

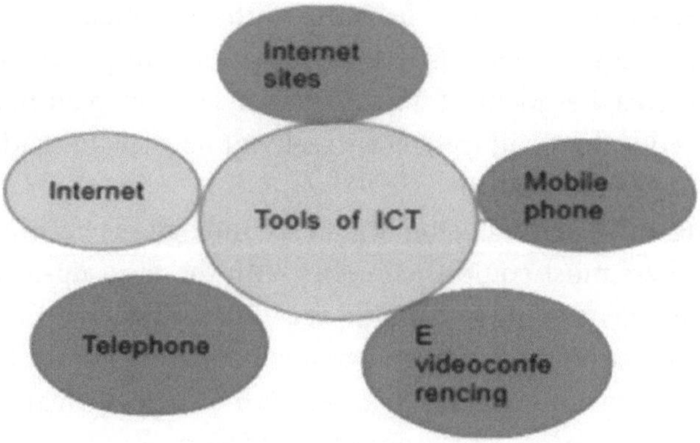

FIGURE 6.2 The stages of teaching and learning using ICT.

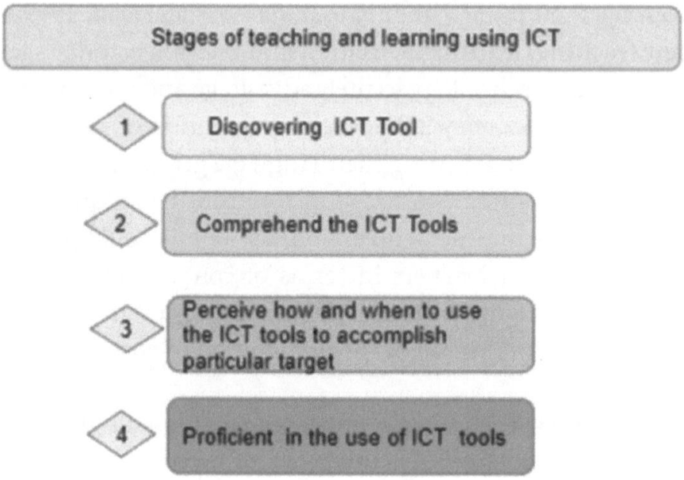

FIGURE 6.3 Types of ICT tools and techniques.

6.4.1 Telephone

The telephone is the most widely used communication instrument on the planet. It is the first electronic device that was used to transmit information and ideas with people and specialists all over the world.

6.4.2 Mobile Phones

Learners prefer mobile handset technology as one of the developed communication systems to access the benefit at any time and from any location. It now serves a critical function in disseminating information to others. Learners can access e-mail and other network benefits directly from their mobile handset, allowing them to collect all education-related information from any location (Jonassen 1991).

6.4.3 Internet

The internet is a successful human mindset invention. The internet shrinks the globe, allowing learners to quick access to relevant knowledge. It is a watershed moment in the history of communication technology. To chart and communicate with everyone without failing performance, e-mail is preferred. It is simple to exchange education-related information with others, and it allows learners to obtain information on their own time.

6.4.4 Internet Sites

Learners can now obtain education-related information through websites. The websites are created by professionals, who also post and update the material. Even schoolchildren are now taught to use websites for educational purposes. Teachers' workload is reduced, and learners are guided to become self-learners and motivators, thanks to websites (Ahmad 2012).

6.4.5 Video Conferencing

Video conferencing is a technique that combines video and speech communication to allow people to connect with each other through the internet. To use this feature, the user will need a webcam, microphone, and a broadband connection. By avoiding the physical appearance for face-to-face interaction, learners can profit from real-time natural conversation.

In educational institutions, information and communication technology is used in the following ways.

6.4.5.1 Interactive White Boards

Interactive boards are a type of advanced information technology used in classrooms. Teachers are able to construct engaging lessons with intriguing content thanks to information technology (Rana et al. 2020).

6.4.5.2 Simple Conduction of Experiments

The use of information and communication technologies improves experiments and makes them more visible. Computers are preferable for obtaining precise results from experiments, and some experiments are impossible to do without them.

Support for audio and video: Students are encouraged to do in-depth analyses by presenting the preparation with audio and video effects, and they can put forth a genuine effort in their preparation (Samuel et al. 2006).

6.4.5.3 PowerPoint

PowerPoint is the most effective tool for faculty members to use in their preparation. Photographs of the subject matter can be brought directly to the classroom by the teachers. Students can also clearly see the teachers' individual preparation. Graphics on the computer images are used to demonstrate and convey the concept of a specific item. The use of computer graphics helps to reinvent creativity. The usage of computer graphics by teachers is encouraged by information technology (Papadakis et al. 2018).

E-mail is used to keep the students in touch on a regular basis. The interaction between students and teachers creates a good learning environment. Students submit their homework to their teachers and receive feedback from them electronically.

6.4.5.4 Putting up Educational Materials

Students can access the instructional materials using specialized software developed by the university. The students gather the required materials from the software. Attitude shifts among teachers. In teaching practice, innovative methods to teaching make a difference (Park et al. 2012). With the use of information technology, teachers can readily refresh their expertise while fulfilling their commitments. It causes a shift in the teacher's mindset (Cocea et al. 2006).

6.4.6 Various Kinds of ICT Tools

At all stages of school, the computer is the most frequently used ICT instrument (Cronin 2002). Changes in computers have been brought about by current developments and advancements in ICT (Day et al. 2004). Smart phones and laptop computers are now readily available to both professors and pupils. Teachers and students in higher education employ computer, laptop, and mobile technology to integrate computer technology into the learning process (Mullamaa 2010). The advantages of using computers in higher education include easy access to helpful material within the classroom, up-to-date and advanced knowledge of relevant course content, and learners' active participation in the learning process (Dornyei and Ottó 1998).

6.5 DIGITAL REVOLUTION IN EARLY CHILDHOOD EDUCATION

What a youngster goes through in their initial years of life has a long-term impact on the development of their minds and the thought process. While genes lay the foundation for our sentiments, vocabulary, motor coordination, sights, and memories, experience also builds the neurological connections that lay the foundations for our emotional responses, linguistic skills, motor skills, visions, and happy times (Gardner 2011). When a particular area receives more simulation, the neuronal pathways within those particular regions, as well as the connections of the other surrounding neurons, become stronger. 'Synaptic pruning' is the term for this process. Imagine your brain as a planet. The motor coordination metropolitan area, memorization mountain, and perception village are among the highlights (Gee 2008). Famous cities have grown in size over time, as have the roadways connecting them. There could be traffic congestion if one region's highways were never established. As a result, the entire brain's growth is slowed. When it comes to getting to work, some neurotransmitters will be exhausted (Gottlieb 2008). Others will take shortcuts and get lost. Betty Hart and Todd Risley investigated children's linguistic exposure. On average, 600 words per hour were presented to youngsters from families on welfare. Children from wealthy families received approximately 2000. By the age of three, the difference had grown to 30 million words. But it didn't stop there: children from wealthy families received far more positive evaluations. There was only one word of disappointment for every six words of praise. Children on welfare, on the other hand, have been subjected to twice as many discouragements as praise This has the potential to make a lifelong difference. Preschool education not only provides academic nourishment to children, but it also provides them with interpersonal skills, courage, and tenacity. When this mix of strength of character is observed it is often known as grit, which is ultimately accountable for the achievements of the entire life span (Handahl 2004). The Heckman curve was first published in 2006 by Nobel laureate economist James Heckman, and it depicts the return on investment in education, which is strongest in the early periods. Since then, the government has begun to take action. Parents in Germany receive a lot of economic assistance to help them raise their children. In Japan, both parents are eligible for one year of parental paid leave. In France, kindergarten is provided free of charge to all children (Martens et al. 2004; Kiggins and Cambourne 2007).

Early Child Education (ECE) comprises education and childcare services provided for children from birth until the age of eight. ECE is frequently concerned with assisting children in learning through play. For most parents, childcare and education are inextricably linked To promote and enable a greater educational component in the caregiving field, collaboration between schools, childcare, and health services should be encouraged. Teachers who make decisions based on knowledge and training of young children are known as ECE professionals. These professionals strive to be lifelong learners in the field they have chosen and are committed to meeting the care and educational needs of their students. There are different early childhood educator terms presented mainly by europeon and Asian countries such as the Child Act 2001, Child Class Centre Act 308, Early Childhood Care and Education Act 1996, and others, which must be defined and understood by early educational leaders. Among the various effects of ECE, the following are the major ones:

1. Essential in fostering school readiness.
2. Correlated to long-term success in school and life.
3. Effective crime prevention strategy.
4. Closes the achievement gap between 'at-risk' students and their peers.

Five facts about using ICT in early childhood education:

1. ICT provides opportunities for children to enhance their learning.
2. ICT elaborates and extends children's creativity.
3. ICT is responsible for development of communication and early literacy skills.
4. ICT enables children to collaborate and work, which further develops social and behavioral skills.
5. ICT promotes analytical skills including problem-solving, experimenting, inquiry, doubt-solving, asking questions without hesitation and decision-making.

Today's early childhood teachers have access to a vast array of ICT tools and materials. As a result, it's critical to make the right decisions when it comes to selecting or picking helpful ICT solutions for early childhood education. When it comes to incorporating ICT in educational experiences, however, there are numerous factors to take into account (Townsend and Bates 2007).

6.5.1 Meaningful ICT Resources and Tools in Early Childhood Education

The first consideration for meaningful ICT resources is the developmental appropriateness. When it comes to making these resources meaningful, this is the most significant. They need to meet a criterion that the ICT tool should be educational, encourage collaboration, support the integration throughout the curriculum, and be used in key areas such as numeracy and literacy (Sonja and Samuelsson 2003). Research also shows that the early ICT resources could be software and/or hardware as long as it and challenges students intellectually and develops higher-order thinking skills, such as decision-making, especially when it comes to developing child ICT capability. The criteria also states that it needs to be transparent and intuitive and avoid any violence or stereotyping. While children are learning to use ICT through their curriculum and when they role play, they should be made aware about safety and various health issues due to the use of ICT in their lives (Preradovic et al. 2016)The second half of this dimension is that once you've chosen your ICT tools and met these requirements, they won't be genuinely useful unless you follow them up with Information based consequences pedagogical approaches. This is where a vital role is performed by ECE professionals. In 2010, UNESCO reinforced the importance of early intervention to introduce ICT skills in the classroom. According to research, simply being exposed to ICT techniques and resources in the classroom studies is no longer adequate for developing ICT skills.

As a result, the facilitator should first adopt formative assessment methods and be able to utilize their expertise in observing children employing ICT in role play, planned learning, including in learning. It is also important that the facilitator should understand how progression in ICT capabilities work and what they can do to facilitate it and one must not be forced to get it about (Furman Melina et al. 2019).

Various ICT technologies used in early childhood education and development are:

1. Talking word processor
2. Drawing and painting programs
3. Digital cameras
4. Video and audio recorders
5. Talking books

6.6 ASSISTIVE TECHNOLOGY FOR ECE

When parents of young children think of the phrase 'assistive technology' (AT), the three words that come to mind are personal computers, mobile phones, and tablets. 'My youngster is too young for computers and tablets', is a common objection of many parents. The truth is that assistive technology refers to a broad variety of devices that include but are not limited to laptops and tablets. Assistive technologies are a form of technology that may be used to help newborns, toddlers, and preschoolers with a variety of needs. Some of them we can purchase, some we can make, and some we can modify. It is entirely dependent on the child's requirements. These are just a few examples of assistive technologies: the bottle

TABLE 6.1 Assistive Technology Tools for the Classroom

S. No.	Tool
1.	Word prediction
2.	Screen readers
3.	Sip and puff systems
4.	Video magnifiers
5.	Word talk
6.	Face mouse
7.	Supernova access suit
8.	Closed captioning and subtitling

holder encourages babies to grasp bottles independently; the grow clock assists young children in learning when it's time to go to bed and when it's appropriate to get out of bed; a slanted clipboard makes it easier for kids to practice art or write a story; perfectly matched skills and sounds are integrated in sound boxes[63] when the preschooler matches a sound-producing block to another sound-producing block. Yes, AT involves the use of laptops and tablets with numerous monitoring tools, applications, and programs that, when appropriately matched to the needs of the children, make them excellent tools (Parette and Stoner 2008). Some of the important assistive technology tools used in classrooms are listed in Table 6.1.

6.6.1 Myths about Assistive Technologies

There are lots of myths about assistive technologies. Assistive technologies include dictation or text speech or audio books. Various myths associated with AT are:

1. 'Using assistive technology is cheating' is a myth. The most common misconception is that using AT is cheating, which is nonsensical because you wouldn't accuse a child with a medical condition that requires them to wear eyeglasses of dishonesty; and you wouldn't accuse a child who is in a wheelchair due to mobility impairments of cheating. In conclusion, it's important to raise awareness that assistive technologies are tools that students with concentration and attentiveness challenges require in order to succeed in school and achieve their academic goals.

2. Assistive technologies lower potential. Another myth is that when kids use AT, they somehow will become over-reliant on it, or they will not be giving their best effort in school. When students are able to use AT, they become more enthusiastic because they have techniques that allow them to accomplish their goals, and they are then able to offer their all since they have tasted success.

3. Use of audiobooks can make it difficult for the child to read. Many guardians are worried that if they allow their child to listen and learn from audiobooks, the kid might not be able to read properly. But many experts say that listening to audiobooks will make children a better reader since reading and listening at the same time with the help of audio books makes a multidisciplinary reading expert.

4. Assistive technology could be learned by learners on their own. The thinking of people sometimes varies: they think that children who are digitally and technologically native have the expertise to teach themselves about the use of assistive technologies. However, just because a child is a digital native born in a time when everybody has a tablet or a smartphone, sends text messages, and watches videos on YouTube, they still may need support and training so that they can make use of these technologies in completing their school projects and homework with more efficiency.

5. Assistive technology gives a biased desirability. Another myth is when kids use AT, they have an unfair advantage over kids in the class who aren't using them. They believe that every child might benefit from assistive technologies such as dictation, audio books, or text speech. It can enable pupils who don't have a learning or attention problem to become much more creative in school. It's a tool that allows individuals with learning and attention challenges to accomplish their goals. As a result, all children in the classroom will be able to use the same assistive technology tool, and everyone will prosper (Burne et al. 2011).

Various assistive technology used for literacy:

1. Text to speech (TTS)
2. Assistive technology for teaching
3. Assistive technology for writing
4. Dictation technology (speech to text)
5. Digital literacy practices in an inclusive classroom
6. Incorporating technology into literacy instructions
7. AEM Navigator
8. TechMatrix

6.7 COMPARISON OF TRADITIONAL PEDAGOGY AND EMERGING PEDAGOGY ENABLED BY ICTS

Pedagogy is the method and practice of teaching. Andragogy is a new method and refers to the practice of teaching adult learners. The word 'pedagogue' comes from the Greek language and means 'child leader'. Pedagogy is frequently employed in education, and it is also used in ways that are not well understood. It could be called an art to teach and help kids to perform various tasks. However, the term currently has a very particular meaning that refers to the construction of an instructional process that results in sharing knowledge to an independent learner. Pedagogy is not about putting some videos, putting some inserts and putting some assessments and saying 'I have created a pedagogy'. However, it is a technique that one establishes for pupils so that they may take a task that they are attempting to understand or a different subject matter that they are attempting to learn and then learn this new procedure through that process. Pedagogy begins with information, then moves on to how that knowledge is presented in an individualized learning strategy, then connects to the assets that must be included in that composition. Finally, those resources are put together to create an intellectual psychology-based educational process, with some assessment process at the end to ensure that the learning process is working properly. So, in a nutshell, pedagogy is the theory, philosophy and methods of teaching, and, strictly, it refers to the teaching of children (Mikre, Fisseha 2011).

Traditional education, also known as 'back to basics', conventional education, or traditional education, was one of society's first declared rules in schools. More holistic approaches that focus on individual student needs, comprehension, and self-control are some forms of education changing more for the adoption of progressive education practices. Traditional teaching methods that emphasize learning and memorization, according to reformers, must be abandoned in favor of student-centered and task-based learning approaches. Many conservative individuals and parents, on the other hand, are worried about the preservation of the goal of objective training standards based on testing, which prefers a more traditional methodology. Depending on the setting, progressive or modern education, developmental psychology-based education, or alternative education may be the antithesis of traditional education. Traditional education is defined differently depending on geography and historical periods (Table 6.2). Traditional education's main goal is to pass on to the next generation the skills, information, and conceptions of current and social conduct that are seen to be essential for the future generation's material and social success (Webb and Cox 2004).

Teachers are the gateways *via* which this knowledge is conveyed, and these norms of conduct are perpetuated. The primary educational tactic in traditional education has been simple oral recitation – a common approach. Students sat at their particular places and listened to one student after another reciting lessons from the book until each had given a chance to read from the book. The major responsibility of the teacher was to assign the particular paragraphs to students and carefully listen to these recitations, correcting individuals if reciting wrong. Then students are required to read, study, and memorize the same chapters given for the assignments to be completed at home. A quiz or assessment may be given at the end of the unit, and the coursework preparation recitation test cycle will continue. It was also a tremendously inefficient use of student and instructor time, with an over-emphasis on verbal replies, sole reliance on memorization, memorizing with little grasp of meaning, and memorization divorced from related assignments. This technique

TABLE 6.2 Major Lines of Difference between the Emerging and Traditional Pedagogies

S. No.	Emerging pedagogy enabled by ICTs	Traditional pedagogy
1.	Activities defined by learners	Activities prearranged by teacher
2.	Waged in teams	Individual
3.	Integrating theory and practice	No connection between theory and practice
4.	Find novel solutions to problems	Apply identified solution to problems
5.	Diagnostic	Summative
6.	Student-centric	Teacher-centric

also required that all students be taught the same content and information, and that pupils who did not learn rapidly be punished rather than applauded for progressing at their own pace (Cabrera et al. 2002).

The influence of ICTs on education can be approached in various ways to meet the necessity of novices. Some important ones are:

1. Active teaching and learning approaches
2. Collaborative teaching and learning approaches
3. Creative teaching and learning approaches
4. Integrative teaching and learning approaches
5. Evaluation teaching and learning approaches

6.8 CONCLUSION

ICT is not only enriching the learning environment of young children but possesses a perceptible enhancement in their overall development including language development, mathematical thinking, ability to express their ideas, views, and thinking, information literacy, reasoning, and active exploration of their physical world around them. ICT as a medium for education is becoming more and more accepted these days and especially during the recent pandemic time. The above chapter has summarized some very positive aspects of ICT over the traditional teaching methods and its stimulation over students as well as teachers. With the invention of these fascinating, attractive ICT tools, even very young children are getting endless opportunities for expression and collaborative learning concepts. The rigid borderline of traditional teaching is becoming blurred. But with all the positive aspects, still many parents need to be counseled on the positive aspects of introducing ICT in their young child's education and resolve the traditional related myths. The pedagogy needs to be explored more for student satisfaction and learning quality. Learners' needs and preferences should be monitored rather than just providing a web-based environment for learning. A student-centered approach that is based on modernization, which enhances student passion and engagement, is likely to be helpful in this regard.

REFERENCES

Ahmad, J. (2012). English language teaching (ELT) and integration of media technology. *Procedia-Social and Behavioral Sciences*, *47*, 924–929.

Al-Jarf, R. S. (2004). The effects of Web-Based learning on struggling EFL college writers. *Foreign Language Annals*, *37*(1), 49–57

Altun, M. *The integration of technology into foreign language teaching International Journal on New Trends in Education and Their Implications* 6, *1* (2015): 29–34.

Atas (2019). Amine Hatun, and Berkan Çelik. *Malaysian Online Journal of Educational Technology.* "Smartphone use of university students: Patterns, purposes, and situations.", *7*(2), 59–70.

Borthwick, K., & Gallagher-Brett, A. (2014). 'Inspiration, ideas, encouragement': Teacher development and improved use of technology in language teaching through open educational practice. *Computer Assisted Language Learning*, *27*(2), 163–183.

Burne, B., Knafelc, V., Melonis, M., & Heyn, P. C. (2011). The use and application of assistive technology to promote literacy in early childhood: A systematic review. *Disability and Rehabilitation: Assistive Technology, 6*(3), 207–213.

Cabrera, A. F., Crissman, J. L., Bernal, E. M., Nora, A., Terenzini, P. T., & Pascarella, E. T. (2002). Collaborative learning: Its impact on college students' development and diversity. *Journal of College Student Development, 43*(1), 20–34.

Chenoweth, N. A., Ushida, E., & Murday, K. (2006). Student learning in hybrid French and Spanish courses: An overview of language online. *CALICO Journal*, 115–146.

Chun, D. M. *Computer applications in second language acquisition: Foundations for teaching, testing and research* (2003): 337–339.

Chun, D., Kern, R., & Smith, B. (2016). Technology in language use, language teaching, and language learning. *Modern Language Journal, 100*(S1), 64–80.

Clark, R. C., & Mayer, R. E. (2016). *E-learning and the science of instruction: Proven guidelines for consumers and designers of multimedia learning.* Chichester: john Wiley & sons.

Cocea, M., & Weibelzahl, S. (2006). Motivation–included or excluded from e-learning. In *CELDA* International Conference on Cognition and Exploratory Learning in Digital Age, *2006* (pp. 435–437). Lisbon: IADIS.

Cohen, L. (1976). *Educational research in classrooms and schools: A manual of materials and methods.* London; New York: Harper & Row.

Collis, B., & Jung, I. (2004). Uses of information and communication technologies in teacher education. In *Teacher education through open and distance learning* (pp. 187–208). Routledge.

Cronin, M. (2002). The empire talks back: Orality, heteronomy and the cultural turn in interpreting studies. *Interpreting Studies Reader*, 386–397.

Day, C., & Sachs, J. (2004). Professionalism, performativity and empowerment: Discourses in the politics, policies and purposes of continuing professional development *International handbook on the continuing professional development of teachers, 3*, 32.

de Kadt, E., Junne, G., Ndiaye, A., & Spiro, E. (2001). *The international institute of communication and development (IICD).*

Dhital, H. (2018). Opportunities and challenges to use ICT in government school education of Nepal. *International Journal of Innovative Research in Computer and Communication Engineering, 6*(4), 3215–3220.

Dornyei, Z., & Ottó, I. *Motivation in action: A process model of L2 motivation* (1998).

Dudeney, G. (2000). *The Internet and the language classroom.* Cambridge: Cambridge University Press.

Floris, F. D. (2014). Using information and communication technology (ICT) to enhance language teaching & learning: An interview with Dr. A. Gumawang Jati. *TEFLIN Journal, 25*(2), 139–146.

Furman, M., De Angelis, S., Dominguez Prost, E., & Taylor, I. (2019). Tablets as an educational tool for enhancing preschool science. *International Journal of Early Years Education, 27*(1), 6–19.

Gage, J. (2013). *How to use an interactive whiteboard really effectively in your secondary classroom.* David Fulton Publishers.

Gardner, H. E. (2011). *The unschooled mind: How children think and how schools should teach.* Basic books.

Gee, J. P. Är videospel bra vid inlärning? *Are videogames good for knowledge acquisition? Kulturens Studia Generalia. Svenska kulturfonden. Oy Nord Print* (2008): 8–23.

Gimenez, J. (2000). A self-access center for business English learners: Meeting individual learners' needs. In *English for Specific Purposes*(24), 37–44. TESOL.

Gottlieb, H. (2008). Digitala pedagogiska artefakter–nya verktyg för pedagoger och intendenter./ Digital artefacts–new tools for teachers and museum personnell. *Kulturens Studia Generalia*, 25–38.

Green, A. (2007). Washback to learning outcomes: A comparative study of IELTS preparation and university pre-sessional language courses. *Assessment in Education, 14*(1), 75–97.

Handahl, G. "Ethics of Teaching in Higher Education." *Teacher trainer courses: Teaching in Higher Education given at Tartu University. Oslo university-Tartu university.* Tartu: University of Tartu 2007 (2004).

Hashim, Y. (2010). Determining sufficiency of sample size in management survey research activities. *International Journal of Organisational Management and Entreprenuership Development,* 6, 119–130.

Hennessy, S., Ruthven, K., & Brindley, S. U. E. (2005). Teacher perspectives on integrating ICT into subject teaching: Commitment, constraints, caution, and change. *Journal of Curriculum Studies, 37*(2), 155–192.

Jonassen, D. H. (1991). Hypertext as instructional design. *Educational Technology Research and Development, 39*(1), 83–92.

Kessler, G. (2018). Technology and the future of language teaching. *Foreign Language Annals, 51*(1), 205–218.

Kiggins, J., & Cambourne, B. The knowledge building community program: A partnership for progress in teacher education. *Handbook of teacher education.* Dordrecht: Springer (2007). 365–380.

Laudari, S. (2019). *"Breaking barriers: Exploring digital practices of teacher educators in Nepal."* [PhD Diss.].

Loveless, A. (2011). Technology, pedagogy and education: Reflections on the accomplishment of what teachers know, do and believe in a digital age. *Technology, Pedagogy and Education, 20*(3), 301–316.

Mahnke, F. H. (1996). *Color, environment, and human response: An interdisciplinary understanding of color and its use as a beneficial element in the design of the architectural environment.* Chichester: John Wiley & Sons.

Markee, N. (1997). *Managing curricular innovation, 198.* Cambridge: Cambridge University Press.

Martens, R., Gulikers, J., & Bastiaens, T. L (2004), Judith Gulikers, and Theo Bastiaens. "The impact of intrinsic motivation on e-learning in authentic computer tasks.". *Journal of Computer Assisted Learning, 20*(5), 368–376.

Mikre, F. (2011). The roles of information communication technologies in education: Review article with emphasis to the computer and internet. *Ethiopian Journal of Education and Sciences, 6*(2), 109–126.

Molnar, G. (2007). New ICT tools in education–Classroom of the future project. In The fourth international conference on informatics, educational technology and new media in education. *AD Novi Sad* (pp. 332–339).

Mullamaa, K. (2010). ICT in language learning--Benefits and methodological implications (international edn.)*ucation studies, 3*(1), 38–44.

Neo, M., & Neo, K. T. K. (2001). Innovative teaching: Using multimedia in a problem-based learning environment. *Journal of Educational Technology and Society, 4*(4), 19–31.

Papadakis, S., Kalogiannakis, M., & Zaranis, N. (2018). The effectiveness of computer and tablet assisted intervention in early childhood students' understanding of numbers. An empirical study conducted in Greece. *Education and Information Technologies, 23*(5), 1849–1871.

Parette, H. P., & Stoner, J. B. (2008). Benefits of assistive technology user groups for early childhood education professionals. *Early Childhood Education Journal, 35*(4), 313–319.

Park, N., & Ko, Y. (2012). Computer education's teaching-learning methods using educational programming language based on STEAM education IFIP international conference on network and parallel computing. Berlin, Heidelberg: Springer.

Pearson, J. (2003). Information and communications technologies and teacher education in Australia. *Technology, Pedagogy and Education, 12*(1), 39–58.

Picciano, A. G. (2002). Educational leadership and planning for technology. *Prentice hall, Inc., A Pearson education company, one lake street, Upper Saddle River, NJ, 07458.*

Preradovic, N. (2016). Mikelic, Gordana Lesin, and Damir Boras. *Introduction of digital storytelling in preschool education: A case study from Croatia. Digital Education Review*, 94–105.

Pun, M. (2013). The use of multimedia technology in English language teaching: A global perspective. *Crossing the Border: International Journal of Interdisciplinary Studies*, *1*(1), 29–38.

Rana, K., Greenwood, J., & Fox-Turnbull, W. (2020). Implementation of Nepal's education policy in ICT: Examining current practice through an ecological model. *Electronic Journal of Information Systems in Developing Countries*, *86*(2), e12118.

Samuel, R., & Bakar, Z. (2006). The utilization and integration of ICT tools in promoting English language teaching and learning: Reflections from English option teachers in Kuala Langat District, Malaysia. *International Journal of Education and Development Using ICT*, *2*(2), 4–14.

Sheridan, S., & Samuelsson, I. P. (2003). Learning through ICT in Swedish early childhood education from a pedagogical perspective of quality. *Childhood Education*, *79*(5), 276–282.

Stepp-Greany, J. (2002). Student perceptions on language learning in a technological environment: Implications for the new millennium. *Language Learning and Technology*, *6*(1), 165–180.

Tinio, V. (2002). ICT in education. *Preliminary Findings*, 6–8.

Townsend, T., & Bates, R. (2007). *Handbook of teacher education: Globalisation, standards and professionalism in times of change*. Berlin: Springer.

Vijiyarani, S., & Suganya, E. (2015). Research issues in web mining. *International Journal of Computer-Aided Technologies*, *2*(3), 55–64.

Webb, M., & Cox, M. (2004). A review of pedagogy related to information and communications technology. *Technology, Pedagogy and Education*, *13*(3), 235–286.

Wishart, J., & Blease, D. (1999). Theories underlying perceived changes in teaching and learning after installing a computer network in a secondary school. *British Journal of Educational Technology*, *30*(1), 25–41.

Yunus, M. M. (2007). Malaysian ESL teachers' use of ICT in their classrooms: Expectations and realities. *ReCALL*, *19*(1), 79–95.

Yunus, M. Md., Lubis, Maimun A., Lin, C. P., & Wekke, I. S. (2009). Language learning via ICT: Students' experience. In *Proceedings of the 5th WSEAS/IASME international conference on Educational Technologies (EDUTE'09)* (pp. 136–142).

Zain, M. Z. M., Atan, H., & Idrus, R. M. (2004). The impact of information and communication technology (ICT) on the management practices of Malaysian Smart Schools. *International Journal of Educational Development*, *24*(2), 201–211.

CHAPTER 7

Conceptual Framework of Digital Transformation in Higher Education an Emerging Trend

Kuldip Kaur and Priyanka Chhibber

CONTENTS

7.1 INTRODUCTION

Technology is transforming life everywhere. The Fourth Industrial Revolution is the wave of technological innovation. The challenge of this Industry 4.0 is to prepare new policies and strategies for human resources so that people can compete and meet the various technological needs of industry. It is necessary for higher education institutions to pay

DOI: 10.1201/9781003254942-7

attention and produce graduates ready to face the challenges of rapid industrial development. The digital transformation has impacted higher educational institutions drastically as institutions have started to adopt digital tools like LMS, an online platform for academicians, students, and administrators that provides distance learning, MOOC (that is massive online open courses), and other resources. Educators and students need not meet physically; rather they are provided online education platforms where they have access to e-learning resources including books, libraries, forums, journals, etc. The students are placed in the middle of a learning system which creates a global learning environment for teaching and learning. In this competitive scenario, the digital transformation needs new educational models in order to bring technological change in organizations' strategic directions and value propositions. New digitized technologies like artificial intelligence, virtual reality, cloud computing, block chain, robotics, Internet of Things (IoT), big data, etc. have produced a powerful disruptive effect in almost every area. Digital transformation creates new opportunities and challenges for higher education and research. The current study provides knowledge related to the need for digital transformation in higher education and specifies its key elements.

7.1.1 Background of Study

The First Industrial Revolution started in the 1700s and has since progressed with step-by-step technological transformation. It involved the introduction of new machines and techniques in agriculture, textiles, iron and steel industries, and the power sector, including steam engines and railroads. The Second Industrial Revolution, which began in the late 19th century and continued until the beginning of 20th century, introduced remarkable advancements in manufacturing and production, such as assembly lines, gasoline engines, railroad networks, telegraph, electricity, etc. The Third Industrial Revolution started in the early 1970s brought drastic changes to the technology and renewable energy sectors. This new era of digitalization brought automation to production and supply through the emergence of electronics, telecommunication, information technology, internet, computers, green electricity, etc. Now the Fourth Industrial Revolution (4IR) also known as 'Industry 4.0' or the 'technological revolution' has brought social transformation through adoption of advanced digital tools and technologies to attain digitalization through increased automation. The emerging technologies used in 4IR bring integration among biological, physical, and digital tools. Figure 7.1 shows the technologies for Industry 4.0.

Table 7.1 highlights the effects of different industrial revolutions on higher education. There are various scientific, technological, economic, social and political challenges before digital transformation under Industry 4.0. The major challenge for workers regards acquiring the advanced skills needed for Industry 4.0. It is mandatory that future workers acquire these skills before the disparity between demand for and supply of skilled labor further intensifies. For universities and higher education institutions, this means immediate adoption of digitalization in order to develop excellence and expertise and to prepare the workforce for Industry 4.0. The learning of theoretical and practical concepts of Industry 4.0 at the university level will definitely help the students in its application at the industry level as universities should deliver courses and training needed for various requirements

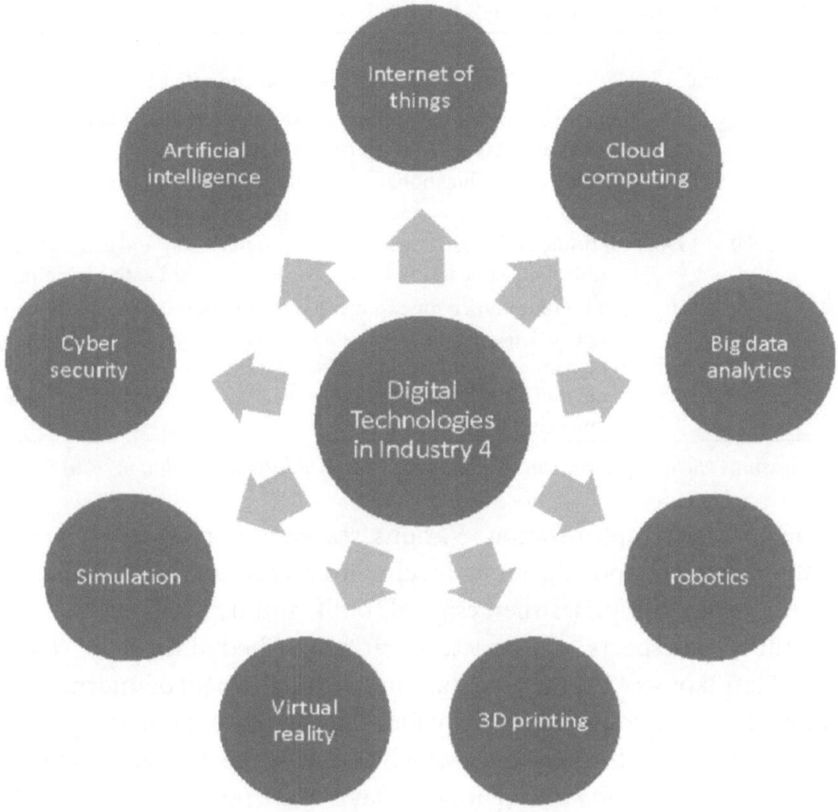

FIGURE 7.1 Digital technologies in Industry 4.0. Source: Selamat et al. (2017).

of Industry 4.0. A well-planned digital transformation in higher educational institutions will ensure a knowledgeable workforce with the right skills and wisdom, but it is not possible without the necessary infrastructure. The purpose of the current study is to identify the need and emerging trends of digitalization in higher education in order to develop a conceptual framework for higher education.

7.2 OBJECTIVES

To identify the need of digital transformation in higher education

To provide a conceptual framework for higher education

To determine various opportunities and challenges for adopting digital tools in higher education under Industry 4.0

7.3 LITERATURE REVIEW

7.3.1 Need for Digital Transformation

A vast literature has been studied to identify the need for digital transformation in higher education. 'Digital transformation' means the integration of digital technologies. It is the process of fundamentally changing an organization from information technology (IT)

TABLE 7.1 Effects of Industrial Revolutions on Education

Industrial Revolutions	Period	Effects on Education
1st Industrial Revolution	Early 1800	Offer various degree options, as well as introducing new education programs to lead educational institutions.
2nd Industrial Revolution	1860 to 1900	Provide new educational opportunities for technicians and engineers and make education accessible for industrial class.
3rd Industrial Revolution	1980 to 1990	It brings web-based technologies, provides easy and extended access to university education, shifting the focus toward active learning pedagogies.
4th Industrial Revolution	After 2000	This era is known for exponential technologies like AI, IOT, computer network technologies, robotics, biotechnologies, etc. In 2012, MOOC courses completely displaced the traditional education system, substantial amendments in curriculum for the rapid development of students in emerging disciplines.

Source: N. W. Gleason, "*Higher Education in the Era of the Fourth Industrial Revolution*" (2018).

modernization to digital optimization. Various researchers give definitions related to technology, digitalization, and educational technology. Lievrouw and Livingstone (2002) stated, 'The technology comprises the designed, built, and deployed artifacts or devices, social and institutional aspects, and structures circumscribed in their use'. Digitalization was defined by Plotnikov as, 'The modern stage in the development of information in which digital technologies are used predominantly for the generation, processing, transmission, storage, and visualization of information which is due to the emergence and dissemination of new hardware and software solutions'. Davis and Hartley (1972) stated that educational technology is, 'The application of physical sciences and engineering technology to provide mechanical instruments or hardware which can be used for instructional purposes and application of scientific principles or software approach to instruction in education and training'. The terms digitization, digitalization, and digital transformation are used interchangeably in the literature. The search with keywords digital transformation in Google scholar and Web of Science resulted in thousands of results. There are various views provided by academicians, government authorities, and business experts on digital transformation.

According to the European Commission, (2017), governments made changes in their mode of operation from manual to digital in order to achieve their objectives of transparency, interoperability, and citizen satisfaction and to improve service delivery.

OECD (2019) provides a roadmap to measure digital transformation through mapping indicators across different areas from education, research, and innovation to trade, social, and economic outcomes.

For the European Union (2021), 'Digital Transformation refers to the adoption of new high impact emerging technologies in businesses including internet of things and artificial intelligence. The European Commission forecasts a transformative, industrial and technological revolution as a global trend till 2030'.

Xu et al. (2018), after studying various opportunities and challenges related to Industry 4.0, states that the Fourth Industrial Revolution is more than just technology-driven change which positively impacts on our core industries like education, health, and business.

Kayembe and Danielle (2019), after considering various opportunities and challenges with regard to Industry 4.0, suggested that the government needs to invest in infrastructure, human capital, and technology in order to develop the education sector to participate in the Fourth Industrial Revolution.

Clark (2018) considers technology as the enabler of digital transformation to enhance, compete, improve customer satisfaction and increase profits through better customer service across the organization's value chain.

Curaj et al. (2018), Technology and digitalization are considered as basic necessities for society, and the universities may provide varied skills and knowledge to learners for a better future which will significantly impact all aspects of their lives.

Rampelt et al. (2019), give the most accurate definition as, 'Digital transformation is a process that influences all activities of higher education institutions and includes the development of new infrastructures, increased use of digital media and technologies for teaching–learning, research, administration, support services, and communication'.

HEA (2019) recommended that in order to move toward digital transformation, every university needs to set up clear, specific, and concrete strategic goals in three main areas: digital infrastructure, development of digital skills of teaching staff and students, and digital culture.

There is a paradigm shift in the way the universities adopt the technologies. Now the interest is focused more on attracting students with advanced knowledge and skills to enhance their experience through competitiveness. In this way, digital transformation becomes a necessity in higher education institutions to improve the teaching and learning.

7.3.2 Skill Requirements for Industry 4.0

In order to know the multiple skills required for industry and ready graduates for their successful corporate careers, there is a need to analyze the current education practices and the industry requirements. According to the World Economic Forum, 'Industry 4.0 will drastically decline the demand for many jobs, including those that require manual skills and physical abilities due to automation of digitalization of the operation process' (Wallner et al., 2016). Table 7.2 highlights the skill requirements for Industry 4.0.

According to Mann et al. (2015): 'There are 9 pillars of digital innovation: autonomous robots, simulation, horizontal and vertical system integration, big data analytics, internet

TABLE 7.2 Change in Skill Requirements in 2022

Trending 2022	**Declining 2022**
Innovation and analytical thinking	Financial management and inventory management
Active learners and emotional intelligence	Installation and maintenance of technology
Creativity, originality, and initiative	Reading, writing, visual, auditory, and speech abilities
Computer-programming	Personnel management and coordination
Complex problem-solving	Quality control, time management, and safety awareness
Leadership and evaluation	
System analysis and social influence	
Cognitive, cross-functional skills	

Source: World Economic Forum 2018.

of things, cyber security, cloud computing, additive manufacturing, and augmented reality. 4IR heavily depends on educational skills which need integration between man and machines'.

According to Xu et al. (2018): The human resources ready for 4IR must possess the specialized skills which are not limited to education alone but cross-functional skills, social competencies, and cognitive skills are also required for holistic development.

According to Lee et al. (2018): in order to meet the challenges and opportunities of digitalization, it is important to understand the complexity and technicality of employment and its requirements according to job specifications and professional competencies according to 4IR.

According to Gottfried and Plasman (2018): A new framework known as 'Career and Technical Education' was developed in place of the traditional academic curriculum in order to respond to the changing demands of the labor market. Tomasello and Brand (2018): There are three mechanisms in CTE framework: *skill building* helps students to gain cognitive thinking for problem-solving and a variety of other skills; *educational engagement* helps in engaging traditional educational content with career-based practical applications through applied conceptual learning and hands-on practice; *Practical experience* helps to transform students' learning skills from traditional (theoretical) to practical (needs-based).

According to Causo et al. (2016): Social robots and intelligent agents are used in education research and innovation to provide support for social intelligence through blended learning and, moreover, they are easy to handle the complex tasks and quite helpful in future education like courses related to STEM.

According to Mourtzis et al. (2018): the higher education institutions use Cyber Physical Systems (CPS) and advanced technologies of Industry 4.0 to create a learning platform where students learn about practical knowledge of manufacturing operations through use of software technologies.

According to Schuster et al. (2016): the knowledge transfer from non-formal learning scenario is more important than from educational institutions alone it means both formal and informal educational structures are needed for knowledge-sharing and, moreover, knowledge is shared easily through technological tools.

It is very much clear from the above literature that all these skill requirements are needed by the future workforce, and for this, higher educational institutions need to be digitally transformed.

7.3.3 Digital Transformation Requires Integration of Higher Educational Institutions with Industry 4.0

The current literature indicates that digital transformation contributes toward integrating higher educational institutions with industry. Digital transformation promotes integration through technology. There are a variety of courses offered by universities at graduate and post graduate levels in different streams such as management, computer science, engineering, technical, and vocational courses in which training and internship programs are part of the course, These trainings and internships are arranged by universities in collaboration with industries, banks, and other organizations. The students get trained during the course in real job conditions, and these projects and programs really help in the integration of universities with industries. Also, students have learned various skills required for industry by the end of their course. Moreover, many industries provide online training and internship programs which also help the students gain knowledge about various skills and competencies.

Gleason (2018): In order to integrate higher educational institutions with Industry 4.0 there is a need for an educational design framework for critical thinking on exponentially growing technologies in which students possess the abilities to improve the sustainability of advanced technologies.

Isenhardt et al. (2015): The development of CPS requires professionals from different fields and their collaboration with experts in order to achieve targeted goals which ultimately required restructuring of higher educational institutions.

Richert et al. (2016): In order to solve various open-ended challenges in collaboration, workspaces should be more advanced and fully virtualized for conducting research by skilled students and experts.

Ward (2017): The higher educational institutions support students in entering into skilled jobs with the help of training, and with the increased supply of a trained workforce, there will be a reduction in costs and increased use of advanced technology.

Mohammed and 'Nell' Watson (2019): technology-enhanced learning (TEL) helps teachers to concentrate on their main duty of conducting research, teaching, and mentoring of students by employing AI techniques. The most widely used TEL system is the Learning Management System (LMS) which is basically a software application that provides educational courses, training programs, and other learning programs through flipped classrooms. It helps instructors to create course content in the form of modules, provide teaching material through video lectures, Power Point presentations, and study notes; track students' progress through customized tests and assessment, which is done automatically by creating quiz assignments.

Chang and Wills (2013): e-learning is more effective when it is given in blended mode through interaction between human beings and technology as in blended or collaborative learning. The teaching–learning is done through offline as well as online mode, which is more beneficial for quick learning.

Elliott et al. (2015): for the professional development of higher educational institutions in Industry 4.0 the institutions adopt a LMS which provides opportunity to deliver training effectively through creation of student groups, mentoring, and support communities.

World Economic Forum (2017) Massive Open Online Courses (MOOC): the combination of traditional education system with MOOC enhances the quality of education in Industry 4.0. As per the World Economic Forum, the 'Global Identity' and 'Education for You' concepts can be applied to MOOC, which means if an idea is globally embraced, it would transform global higher education,

Dong et al. (2016): MOOC platforms can make teaching–learning more effective by enabling teachers to adjust their teaching methodology, and with the help of professional software and pocket labs, bring innovation to engineering education.

Lee et al. (2017): flipped classrooms in teaching human morphology are also based on MOOC courses.

Ellahi et al. (2019): in order to integrate with Industry 4.0, enhancement of material delivery mechanisms is needed for curriculum development. For the execution of these delivery mechanisms, the various systems of higher education institutions such as LMS, Human Resource Management, Financial Management, Research & Innovation and Teaching & Learning require integration with Industry 4.0, which aims to create smart factories and smart cities.

From the above literature study, it is very much clear that the digital transformation of higher education institutions aims to create smart higher educational institutions in order to respond to Industry 4.0, where all the systems are interlinked and integrated. The organizations, both private and public, need to collaborate with educational institutions to reshape the academic curriculum to match with current industry needs.

7.4 RESEARCH METHODOLOGY

This study is based on secondary information collected through various journals, books, and articles from Scopus and Web of Science to synthesize the 4IR and higher education. The main aim of the study is to identify the need of digitization in higher education in order to develop a conceptual framework and also identify various opportunities and challenges for adopting digital transformation in higher education. To achieve the objectives, a vast literature was studied regarding the need of digital transformation, skill requirements for Industry 4.0, and collaboration of higher education institutions with industries, and a conceptual framework is developed on the basis of literature survey. After that, various opportunities and challenges are discussed and recommendations are provided to overcome those challenges.

7.5 RESULTS AND DISCUSSION

From the above discussion it is clear that higher educational institutions must embrace digital transformation in order to make students ready for Industry 4.0. A conceptual model shown in Figure 7.2 is constructed on the basis of a literature survey and indicates that the digital transformation requires students to be multi-skilled, including complex problem-solving skills, decision-making skills, critical thinking and innovation skills, emotional intelligence, etc. There is a need for administrative support in the form of sufficient funding and infrastructure development for the investment in digital technologies, like artificial intelligence, robotics, Cloud computing, big data analytics, and Internet of Things. The students' engagement for these programs and their positive attitude toward learning new skills required the universities to arrange various webinars, seminars, conferences, workshops, symposiums for faculty as well as students to create awareness and knowledge about the need for advanced skills and awareness of the problems and challenges in digitization so that they are ready to meet the needs of Industry 4.0. The higher educational institutions also need industrial collaboration to make the students familiar with handling new challenges under expert guidance and also to get better placements. Another important consideration is internet connectivity, as poor internet connectivity creates hurdles in the

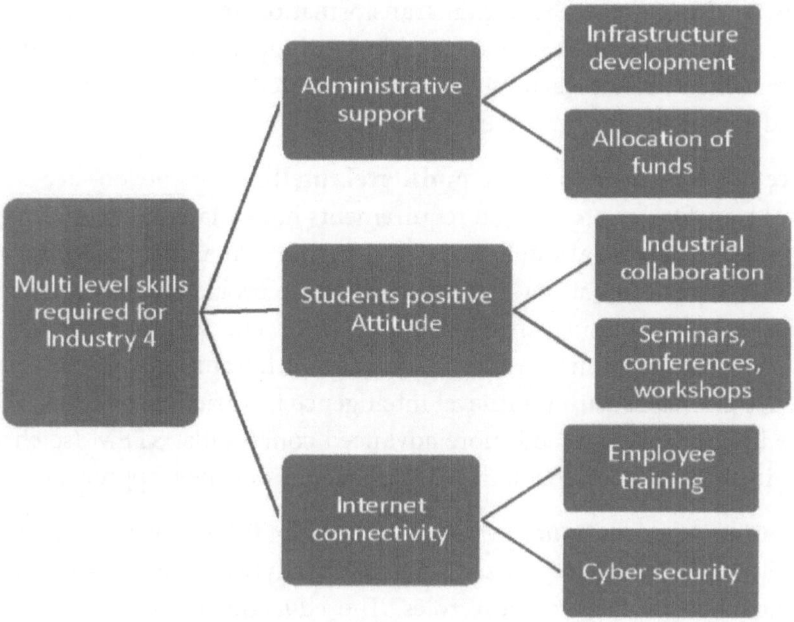

FIGURE 7.2 Conceptual framework for digital transformation of higher education in Industry 4.0. Source: Author.

way of learning. The universities pay special attention to the training of employees to make them digitally literate so that teaching–learning could be made more interactive through blended learning mode, but there are also security issues that the universities need to focus on and resolve in order to be digitally transformed and make the students ready to adapt to the challenges of Industry 4.0.

7.5.1 Challenges for Adopting Digital Transformation in Higher Education

Higher education today faces many challenges regarding lack of funding, change of organization structure, teaching methods, accreditation procedures, restructuring of educational system, multi-skilled graduates and many more which create hurdles in the adoption of digital technologies in higher education:

Lack of Funding: the main challenge preventing higher education institutions from adopting digitalization is lack of funding. Although funding for education has increased over the years, it is not sufficient for complete functioning of higher education and leads to increased university fees and reduced research funding. There is a need of funding more in technological advancement and moreover funding needs to be prioritized. The funds are also needed for providing training to teachers regarding advanced technology and for this a substantial backup is needed.

Change of organization structures: with the emergence of digital transformation there is a change in organization structure which requires change in business models, faculties, departments, and the multidisciplinary cooperative structure.

Change of teaching methods: digital transformation requires changes in teaching–learning methods from classroom teaching to flipped classrooms, online MOOCs, LMS, open and distance learning, and changes in teaching infrastructure, like digital tools and equipment for the virtual world.

To produce holistic graduates with multilevel intelligence: the jobs are changing due to digital transformation. The job requirements have changed from physical or simple computer to artificial intelligence, IoT, big data analytics, robotics, simulation, etc. The skill requirements have totally changed to problem-solving, cognitive skills, analytical decision-making, cross-functional skills, etc., which creates a big challenge for the students who want to work in industries with traditional skills. There is a need to produce graduates with multilevel intelligence in varied disciplines. This requires changes in curriculum to add more advanced courses like STEM (science, technology, engineering, and mathematics), which reconsiders new approaches.

Reorganization of educational systems: Industry 4.0 has created many cross-functional roles which need specialized skills like analytical, problem-solving, and technical know-how to fit into these roles. The educational institutions provide only traditional education especially in humanities and social sciences even in the time of 4IR, hindering the progress of today's labor market. There is a need to restructure the existing higher educational system to train the students with future skills

Integration of lifelong learning: To update the skills of employees throughout their working cycles there is a need for investment in integration and collaboration with businesses and governments or may require partnerships with other stakeholders in order to motivate the individuals. At the university level, funds should be allocated to train the students with industries which will increase the skilled workers in every discipline.

Need of trained mentors: there is a need of trained mentors for teaching advanced technological approaches such as AI, IOT, robotics, big data, etc., which require the efforts at university level to provide them training opportunities by organizing various workshops, faculty development programs, and refresher courses so that they become ready to teach advanced skills to students and increase the accessibility of classrooms worldwide through blended learning.

Lack of digital culture: there is a need to create digital culture and an enabling environment for teachers, learners, and administrators to innovate, imagine, create, and collaborate in order to create a digital culture. For this there is a need for administrative support to bring digital culture into the institutions.

Lack of digital standards, norms, and certification: the big challenge for digital transformation is lack of digital standards, privacy norms, and certification which create barriers for higher education institutions to transform digitally.

Digital security risks: digital transformation requires online integration between several entities which prevent security violations and unauthorized hacking of data.

Cyber security issues pose a great threat to the reputation of universities and safety of students so these cyber issues must be dealt with seriously.

Other challenges: in addition to the above challenges there are also challenges regarding lack of resources, poor internet connectivity, shortage of trained teachers, language-related problems, poor maintenance, digital equipment in need of upgrades, etc.

7.5.2 Opportunities in Digital Transformation

Along with the challenges, the digital transformation brings lots of opportunities for higher educational institutions to learn, innovate, and create.

Creativity, research, and innovation: The digital transformation brings an environment of creativity and innovation. The Industry 4.0 STEM curriculum reconsiders new approaches and emphasizes new collaborative skills.

Fostering partnerships and enhanced vocational skills: The students are the main beneficiaries of the educational ecosystem. They are connected with many other stakeholders to get training in their respective courses and learn the multiple skills needed for Industry 4.0. The students, especially engineering and IT graduates, really do wonders after they are skilled in multitasking. The digital transformation provides an opportunity for collaboration with other stakeholders like government, private companies (such as IBM, Microsoft, etc.) for research and development. The universities benefit by this collaboration, as they get an opportunity to train their students with these partner companies, and students get a chance to learn multilevel skills required for industry.

New and advanced learning: The teachers and students both get a chance to enrich their knowledge with advanced technology and learn multiple skills. There are various open learning sources like MOOC through which both educators and learners can enroll themselves in different online courses offered by various universities, receive certification after completion, and benefit from enhancing their knowledge in various fields. India is considered to be the biggest market for MOOCs after the USA. This provides a great opportunity to avail high-quality learning along with traditional education.

Orchestration of classrooms to support teaching–learning: For the professional development of students, training formats like LMS provide an opportunity to higher educational institutions to effectively deliver training to groups of students, through mentoring and support communities. Course contents are organized through quizzes, puzzles, and grades. The blended classroom learning provides support in teaching–learning for both teachers as well as students.

Digitized/flipped classrooms: The digital transformation has brought a complete revolution through 'flipped' classrooms. Through digital screens students get the same base contents and input which increase student engagement in the classrooms, as classrooms become more interactive and interesting. Interactive boards and digital highlighters are very popular devices related to IoT which help in accelerating and simplifying the learning experience as they receive, acknowledge, and reciprocate the information in an easier manner. The digital scanners help in easy interaction with instructors through digitally transferring text to smart phones. Similarly, digital dashboards help in extracting data in minutes. Instead of spending time on checking papers, a single click on the mouse can provide accurate data and report key performance indicators.

7.6 SUGGESTIONS AND RECOMMENDATIONS

In order to overcome the challenges mentioned above and to grab the opportunities provided by digital transformation the following suggestions need to be considered.

Need of raising awareness: for digitalization of higher education institutions, universities need to create awareness among students regarding the latest changes in the job market, the mission to prepare highly qualified and skilled human resources, and the restructuring of the labor market. This awareness should be promoted through seminars, workshops, and by inviting experts from around the world.

Need of planned integration: there should be a planned network of integration among higher educational institutions and industry stakeholders so that the students become aware of practical challenges of industry and get themselves ready to accept all the challenges through real experiences. The planning ensures financial sustainability and enhances transparency in higher education. In order to promote creativity and innovation for Industry 4.0, innovative training models, programs and methods, and start-up educational projects should be encouraged to connect all stakeholders, instructors, learners, lecture halls, labs, etc.

Need of personalized training programs: There should be training programs as per the needs of every learner. Also, curriculum needs to be changed and updated in order to promote research and training of new skills like working with robots, using block chain technology, big data, IoT, artificial intelligence, problem-solving, critical thinking etc.

Need to combine traditional methods with new online methods: the online teaching platforms, like MOOCs and STEM education, should be encouraged so that learners can acquire knowledge through a blended mode of learning and get benefited in their careers. For this there is a need to accelerate the digital transformation process which takes place through planning, by formulating strategies, implementing innovation, and monitoring the impact of technology deployment.

Need of support from top management: in order to update and enhance the knowledge of teachers, improve vocational skills, and teaching methods, priority should be given to placing university lecturers at enterprises to enhance interaction among universities and businesses. The top management should encourage and provide support for international cooperation activities and integration in training. The universities also need to pay attention to digital security and risk for the smooth international connectivity.

Need of research funding: universities should prioritize investment in research projects aligned with Industry 4.0, which requires the allotment of sufficient funds for digital tools and the training of teachers. In order to transform the traditional system into a smart system, universities and government should come forward, making investments in infrastructure as this investment demands huge funds. Universities can arrange funds from various funding agencies and government or UGC grants can also be utilized by setting priorities of investment.

7.6.1 Government Initiatives

The government has from time to time taken various initiatives to support higher educational institutions. One of these initiatives is the creation of the National Education Technology Forum. The motivation for this forum was to provide direction, investment support, and advice to institutions for the success of the Academic Bank of Credit scheme. This scheme makes the higher education institutions more liberalized by offering flexible credit-based degrees and offers various opportunities to modernize curriculum and advocate reforms toward an active teaching–learning mode.

The other major initiative of MHRD is the National Mission on Education through ICT to provide digital education solutions for everyone so that all can receive quality education. The currently ongoing major initiatives by MHRD are SWAYAM, SWAYAM PRABHA, National Digital Library (NDL), E-Yantra, FOSSE, spoken tutorials, and virtual labs.

SWAYAM: was formally launched by the president of India on 9 July 2017 and offered about 2200 courses. Students can enroll in these courses through the SWAYAM platform. Table 7.3 shows the courses hosted by **SWAYAM** (Study Webs of Active Learning for Young Aspiring Minds) in coordination with nine coordinators which are involved in the preparation of courses to cater to the needs of various learners.

SWAYAM PRABHA: a government initiative launched on 9 July 2017 to provide 34 high-quality educational channels through DTH across the country and deliver the most cost-effective e-education to the students of higher education, school education, and the subscribers of DTH services and dish TV. It meets the needs of lifelong learners through ensuring access to quality educational contents to students located in remote and poor areas.

National Digital Library (NDL): NDL is a national knowledge asset providing free access to national and international digital libraries related to education, research, and innovation through ICT and contains all kinds of digital repositories like books, audio, videos, textbooks, and other online learning materials. This framework has been developed by IIT Kharagpur to provide free access to learners; anybody can register free of cost in NDL.

TABLE 7.3 Courses Hosted by SWAYAM in Coordination with Nine Coordinators

SWAYAM Coordinators	Learners Covered
University Grants Commission (UGC), New Delhi	Offer programs for post graduate students (non-technical education)
Consortium for Educational Communication (CEC), Education center in New Delhi	Offer programs for undergraduates covering varied subjects such as arts, social sciences, applied sciences, and professional courses, etc. (non-technical education)
National Programme on Technology Enhanced Learning (NPTEL) initiated by seven institutes of technology	Offer web-based courses for engineering and physical sciences at graduate and post graduate level
National Council of Educational Research and Training (NCERT), New Delhi	Frame curriculum and offer books for school education
National Institute of Open Schooling (NIOS) under the ministry of education	Providing various vocational, community-oriented programs and elementary courses through open basic education programs
Indira Gandhi National Open University (IGNOU)	Provide education through distance learning to all segments of society
National Institute of Technical Teacher Training and Research (NITTTR), Chandigarh	Provide teacher training to faculty and staff of technical education system
Indian Institute of Management, Bangalore (IIMB)	Offer programs for post graduate, doctoral, executive, and management students
All India Council of Technical Education (AICTE), New Delhi	Offer varied programs to develop technical education and also run various MOOCs courses

Source: Digital Initiatives in Higher Education by MHRD – GNDEC retrieved information from https://gndec.ac.in

E-Yantra: is a robotics-based project on embedded systems assigned to IIT Bombay for effective digital education across science and engineering colleges in India. It helps to provide training to teachers and students through organizing workshops in which they learn by doing or solving real problems with the help of advanced technologies. For the active participation of teachers and students various E-Yantra robotics competitions are arranged as an innovative step in problem-solving. E-Yantra helps colleges in their routine training curriculum through robotics labs/clubs and gives benefits to more than 275 colleges across India.

Virtual labs: the students of various disciplines of science and engineering at undergraduate and post graduate level, and even research scholars, get remote access to virtual laboratories through performing virtual lab experiments at their own pace and benefit by deep understanding in their field of study. It provides a complete LMS for teachers and students to learn about various virtual tools with the help of video lectures, live demonstrations through animations, and additional web resources.

FOSSE (Free/Libre and Open-Source Software in Education): This free and open-source software implemented by IIT Bombay is part of a national mission on education which helps to improve the quality of education in the country through promotion of FLOSS tools in education and research, and it is very cost-effective which reduces the dependency on proprietary software in educational institutions.

7.6.2 Initiatives Desired at Universities/Colleges Faculty

The government has started many initiatives for the digital education of students. Now it is the responsibility of universities to come forward to create awareness among students regarding these initiatives.

Encourage students to take SWAYAM courses: universities can start local chapters in coordination with UGC/NPTEL to help the faculty and students regarding registration and delivery of courses. The faculty can make efforts to motivate students by explaining the benefits of these courses. For instance, these courses allow credit transfer up to 20% for the students who enroll. And their name is shown in the local chapter. The course material can also be used by faculty for offline and online teaching and also faculty to be encouraged to prepare the courses for SWAYAM. Funding support is also provided for the creation of MOOCs courses.

The institutions should install DTH facilities and TVs for SWAYAM PRABHA channels: universities must make efforts to make these channels available to students. The students can be motivated to watch as per their convenience the new contents delivered repeatedly related to various courses according to the schedule which will help students to prepare for competitive exams and to better understand their subjects.

Disseminate the information regarding NDL and virtual labs: dissemination of information to students and faculty is a must, and institutions can motivate them to use the e-resources and virtual lab experiments (which do not require any additional infrastructure set-up) free of cost. The simulation-based experiments can be accessed remotely via the internet.

Take initiative to set up an E-Yantra lab: The higher educational institutions should take initiative to set up E-Yantra labs at the college level. These robotics labs are set up in the colleges to impart effective science and engineering education and training through project-based learning. The college needs to register with a mail request to set up an E-Yantra lab and commit a team of teachers to be trained through a two-day workshop; once the process is completed, the lab is inaugurated online.

7.7 CONCLUSION

The main aim of this study is to know about the need of digital transformation in higher education, develop a conceptual framework, and also identify various opportunities and challenges for digital transformation in higher education. To achieve the objectives, a vast literature was studied and a conceptual framework was developed on the basis of the literature survey. After that, various opportunities and challenges are discussed and recommendations are provided to overcome those challenges. The advanced technologies have changed many industries but higher education is still in the beginning stage of adoption. Hence, it is important to assess and evaluate the digital transformation readiness in the higher educational institutions for the challenges and to further embrace the possibilities it has. In the adoption toward digital transformation, a number of challenges were identified including lack of funding, restructuring of educational system, changes in teaching pedagogies, needed curriculum changes, requirements of various skills, etc. Along with challenges, digital transformation also brings many opportunities for learning, innovation, and creativity, like higher productivity, greater flexibility, blended classrooms, lifelong learning, etc. To ensure effective and immediate reorganization of higher educational institutions, there is a need to follow a strategic approach. Innovation, cultural change, and changes in the mindsets of employers and employees are required for technological change and survival in this competitive environment. In order to be ready for Industry 4.0, we need to build an education system that fosters innovation and creativity to train future workers and their lifelong learning. In this rapidly changing scenario, the higher educational institutions need to rethink and improve their responsiveness toward society. The businesses and stakeholders need to develop human resource strategies toward sustainable development goals through cooperation with universities. A new digital approach is required for teaching–learning, research, and skill development to support digitization like web-based learning platforms, video conferencing tools, MOOCs, STEM education, virtual labs for experiments, and digital infrastructure to support the blended learning process. The government has contributed toward digital transformation through various initiatives like SWAYAM, SWAYAM PRABHA, virtual labs, e-Yantra, FOSSE, NDL, and many other digital platforms to make the education digitized; but universities and colleges also need to take initiative to make these platforms available to students by creating awareness and motivating them to get benefit from these online platforms. Faculty need to take initiative to combine the traditional system with a digital education system to make the teaching–learning more interactive and interesting. For this, universities must ensure proper training to existing teachers, acquire new talent and prepare strategies after consulting the experts, allocate sufficient funds for infrastructure development, and lay down policies to minimize security threats. The trend of digitalization in education is a solution for the challenges posed by Industry 4.0. The integration between education and advanced technology can revolutionize the teaching and learning process and improve the overall quality of higher education.

REFERENCES

Causo, A., G. T. Vo, I. M. Chen, and S. H. Yeo, "Design of robots used as education companions and tutor," *Mech. Mach. Sci.*, vol. 37 (2016) pp. 75–84,

Chang, V., and G. Wills, "A University of Greenwich case study of cloud computing: education as a service," In *E-Logistics and E-Supply Chain Management: Applications for Evolving Business*; Graham, D., Manikas, I., Folinas, D.K., Eds.; IGI Global: Hershey, PA, USA, 2013; pp. 232–253.

Clark, E., "Digital transformation: What is it?" (2018) Retrieved from https://er.educause.edu/articles/2018/5/digital-transformation-what-is-it.

Curaj, A., L. Deca, and R. Pricopie, *European Higher Education Area: The Impact of Past and Future Policies*. Springer: New York, 2018.

Digital Initiatives in Higher Education by MHRD- GNDEC retrieved from https://gndec.ac.in

Dong, F., J. Liu, H. Liu, and S. Dai, "The impact of MOOC E-learning on communication engineering," *Icesame*, vol. 1(2) (2016) pp. 1275–1279.

Ellahi, R.M., M.U.A. Khan, and A. Shah, "Redesigning curriculum in line with industry 4.0," *Proc. Comput. Sci.*, vol. 151 (2019) pp. 699–708.

Elliott, M., N. Rhoades, C. Jackson, and J. Mandernach, "Professional development: Designing initiatives to meet the needs of online faculty," *J. Educ. Online*, vol. 12, no. 1 (2015) pp. 160–188.

European Commission, "Tallinn declaration on E Government," (2017). Retrieved from https://www.eu2017.ee/news/insights/tallinn-declaration-egovernment-ministerialmeeting-during-estonian-presidency

Gleason, N.W., "Higher education in the era of the fourth industrial revolution," *High.Educ. Era Fourth Ind. Revolut.*, vol. 1 (2018) pp. 1–229.

Gottfried, M.A. and J.S. Plasman, "Linking the timing of career and technical education course taking with high school dropout and college-going behavior," *Am. Educ. Res. J.*, vol. 55, no. 2, (2018) pp. 325–361.

Hartley, J., & Davies, I. K. Note-taking: A critical review. *Programmed learning and educational technology*, vol. 15, no. 3, (1978) pp. 207–224

Kayembe, C., and D. Nel, "Challenges and opportunities for education in the Fourth Industrial Revolution," *African Journal of Public Affairs*, vol. 11, no. 3 (2019) pp. 79–94.

Lee, J., C. Lim, and H. Kim, "Development of an instructional design model for flipped learning in higher education," *Educ. Technol. Res. Dev.*, vol. 65, no. 2 (2017) pp. 427–453.

Lee, M.; Yun, J.; Pyka, A.; Won, D.; Kodama, F.; Schiuma, G.; Yan, M.R., How to respond to the Fourth Industrial Revolution, or the Second Information Technology Revolution? Dynamic new combinations between technology, market, and society through open innovation. *J. Open Innov. Technol. Mark. Complex*, vol. 4 (2018) p. 21.

Lievrouw, L.A., and S.M. Livingstone, *Handbook of New Media: Social Shaping and Consequences of ICTs*. London: SAGE, 2002.

Mann, C., & Simons, A. Local emergence and international developments of conservation trading systems: Innovation dynamics and related problems. *Environmental Conservation*, vol. 42, no. 4, 2015 pp. 325–334

Mohammed, P. S. and E. 'Nell' Watson, *Towards Inclusive Education in the Age of Artificial Intelligence: Perspectives, Challenges, and Opportunities*, Singapore: Springer, 2019; pp. 17–37.

Mourtzis, D., E. Vlachou, G. Dimitrakopoulos, and V.Zogopoulos, "Cyber-physical systems and education 4.0: The teaching factory 4.0 concept," *Procedia Manufacturing*, vol. 23, (2018) pp. 129–134.

OECD, *Measuring the Digital Transformation: A Roadmap for the Future*, OECD Publishing: Paris, 2019. https://doi.org/10.1787/9789264311992-en

Rampelt, F., D. Orr, and A Knoth, *Bologna Digital 2020 White Paper on Digitalization in the European Higher Education Area*, Romania: Springer 2019.

Richert, A., M. Shehadeh, L. Plumanns, K. Gros, K. Schuster, and S. Jeschke, "Educating engineers for industry 4.0: Virtual worlds and human-robot-teams," *Proc. IEEE* Glob. Eng.Educ. Conf. no. April, (2016) https://doi.org/10.1109/EDUCON.2016.7474545.

Schuster, K., K. Grob, R. Vossen, A. Richert, and S. Jeschke, "Preparing for Industry 4.0 Collaborative Virtual Learning Environments in Engineering Education," *Eng. Educ.*, vol. 4.0 (2016) pp. 477–487.

Selamat, A., R. A. Alias, S. N. Hikmi, M. Puteh, and S. M. Tapsi, "Higher education 4.0: Current status and readiness in meeting the fourth industrial revolution challenges," *Redesigning Higher Education towards Industry*, vol. 4 (2017) pp. 23–24.

Tomasello, J. and B. Brand, *How ESSA and IDEA Can Support College and Career Readiness for Students With Disabilities Considerations for States*, USA 2018; p. 29. http://www.ccrscenter.org/

Wallner, T., Wagner, G., Costa, Y. J., Pell, A., Lengauer, E., & Halmerbauer, G. Academic Education 4.0. In *International Conference on Education and New Developments* (Vol. 2016, pp. 155–159) (2016, June). Lisbon, Portugal: World Institute for Advanced Research and Science

Ward, S., "The impact of shared leadership on a career and technical education campus," Doctoral dissertation (2017).

World Economic Forum, *The Future of Jobs Report 2018; Insight Report*; World Economic Forum: Geneva, Switzerland, 2018.
Xu, M., J. M. David, and S. H. Kim, "The fourth industrial revolution: Opportunities and challenges," *Int. J. Financ. Res.* vol. 9, no. 2 (2018) pp. 90–95.

CHAPTER 8

Sustainable Development Goal 4

Building an Inclusive Learning Environment for All

Nima Anil and S. Kavitha Maithily

CONTENTS

DOI: 10.1201/9781003254942-8

8.1 INTRODUCTION

The world has witnessed the journey of disabled welfare, which began from hospitals and has moved to inclusion. The education of children with disabilities is now focusing on (United Nations) Sustainable Development Goal (SDG) 4. It emphasizes that regular classrooms with special students are more effective in building an inclusive society in developing countries. Education is indispensable, and education acts as a significant instrument that brings change within the individual and in society – educating differently abled students who need more care and effort. 'Literacy is the ability to identify, understand, interpret, create, communicate and compute, using printed and written materials associated with varying contexts'. is the definition of literacy drafted by United Nations Educational, Scientific, and Cultural Organization (UNESCO, 2019).

8.1.1 Disability

In the Oxford Dictionary,"a physical or mental condition that limits a person's movements, senses, or activities". (https://www.lexico.com/definition/Disability, 2021). The preamble of the Right of Persons with Disabilities (RPWD) Act by the United Nations designates disability as the result of interactions between the disabled and the barriers and attitudes of that society toward them that hinders their full and effective participation in society (https://www.un.org/development/desa/disabilities/convention-on-the-rights-of-persons-with-disabilities/preamble.html, 2021).

According to WHO, discrimination against disabled persons limits their access to social benefits. So, it is evident that disability is a complex phenomenon that reflects people's interaction between their bodies and features of the society in which they live. In the medical model, disability is neither purely medical nor purely social, where people with disabilities may experience problems from their health conditions. A balanced approach toward people with disabilities is needed. More than six hundred million people worldwide live with a disability that causes chronic illness, injuries, violence, infectious diseases, malnutrition and other issues related to poverty (WHO, 2011).

One billion people in the world are disabled, which means 15 percent of the world population is disabled. Children with disabilities are significantly less likely to attend school; according to the census of UNESCO, 90 percent of disabled children are not attending schools (https://en.unesco.org/themes/education, 2021).

8.1.2 Inclusive Education

In 1869, the church missionary society supported Jane Leuport in starting a blind school in Benares. It was the first stone of initiation to inclusive education for the disabled. Inclusive education is the only remedy to promote children with disabilities, especially in developing countries. Inclusive education is an strategy created for uplifting special education to another level. Inclusive education is a core area for revised education acts all over the world because of SDG 4. The education system should respond to the needs of diverse and excluded groups in society. Through inclusive education, the right of all children to learn in a stimulating and quality learning environment can be enforced.

Furthermore, general committees were formed by most of the developed and developing countries and made curricular changes to impart inclusive education (United Nations 2016, para. 7). The tenets of inclusive education are (1) all students have their unique abilities, and inclusive education encourages those abilities for higher education and economic stability in the future; (2) social stigma toward the disabled can change through providing inclusive education; and (3) secure social position for specially-abled persons. Inclusive education is a system that provides mentally, physically, and socially challenged students the same as the general students.

According to the census of 2011, 2.21 percent of the population is disabled in India; among them visual impairment accounts for 18.8 percent, hearing impairment 18.9 percent, muscular impairment 20.3 percent, speech impairment 7.5 percent, mental illness 2.7 percent, mental retardation 5.6 percent, multiple disorder 7.9 percent, and other are 18.4 percent. It was found that 61 percent of children with disabilities are attending education in the age group of 5–19 years, 27 percent are not attending education, and 12 percent have participated in education from an educational institute, i.e., effective implementation of inclusive education in India changed the literacy rate of disabled persons from 2001 to 2011 (National Sample Survey Office, 2016).

8.2 SDGS AND INCLUSIVE EDUCATION

'Ensure inclusive and equitable quality education and promote lifelong learning opportunities for all' is the fourth goal of sustainable development (https://sdgs.un.org/goals/goal4, 2021). United Nations education, scientific and cultural organization has conducted a worldwide movement called 'Education for All' (EFA). The main objective of EFA is to satisfy the population's basic needs in terms of teaching and learning by 2030. Quality education for all includes improving skills to learn objectives of other SDGs and achieve the same by 2030 (Table 8.1).

SDG aims to encourage and promote poor and vulnerable children globally who are excluded from society. So for the restart, this mission needs special and joint effort.

TABLE 8.1 Global Target Indicators for SDG 4

Sl. No.	Target No.	Targets
1.	4.1	4.1.1 Improvement of youth and adults in reading (primary education)
		4.1.2 Improvement in basic mathematics (primary education)
2.	4.2	4.2.1 Proportion of children (under 5years) who are developmentally on track in health, learning, and psychosocial well-being
		4.2.2 Participation rate of children with disabilities in organized learning
3.	4.3	4.3.1 Participation of youth and adults in education
4.	4.4	4.4.1 Proportion of adults with information and communication technology
5.	4.5	Parity indices of Education for All
6.	4.6	The proportion of the population that have at least achieved literacy and numeracy
8.	4. a	The proportion of schools with access to basic facilities
9.	4. b	The volume of official development assistance flow for scholarships
10.	4. c	The proportion of teachers with enough qualifications for an inclusive classroom

Source: SDG 4 Data Book, Global Education Indicators 2019.

According to the index score, Finland is ranked first with 85.90 percent in achieving SDGs and the targets and is followed by Sweden with 85.61 percent; India is in the 120th position with 60.07 percent. The least performer in SDGs achievement is the Central African Republic in 165th position with 38.27 percent. These scores are measured according to the overall progress toward all 17 goals (Sustainable Development Goals, National Indicators Frame work Progress, 2021).

To attain Sustainable Development Goals by 2030, children with disabilities need to be enrolled in regular schools and given equal opportunities for participation and learning; school facilities and campuses should be properly equipped disabled-friendly; professional development programs for the teachers need to be conducted; and a quality learning environment for all children, including children with disabilities, must be provided.

The review of policies and legal frameworks at the international level that focus on the education of children with disabilities states:

- Approximately 155 countries adopted and committed to universal primary education in 1990 with UNESCO, UNICEF, and UN development programs.
- The Salamanca conference in 7 – 10 June 1994 in Spain confirmed education objectives for all by prioritizing the fundamental policy of inclusive education. This conference adopted the Salamanca statement on principle and procedures in special education and a framework for action. The main aim was to proclaim education in the Universal Declaration of Human Rights and is a force to be reframed as 'education for all' by world declaration (United Nations, 1994).
- UNCRDP Act (United Nations Conventions on the Rights of Persons with Disability and Optional Protocol) ensures all aspects of life should be easy for disabled people. In this protocol, there are 50 articles and 18 optional articles. Among those articles, the 24th article is about the education of the disabled and the measures the states parties should consider while providing them education. The person with the disability should have the right to an education without discrimination and equal opportunities for the disabled students should be provided. Inclusion of students in lifelong learning and all levels of the education system should ensure three points: a. should develop potential and sense of dignity and respect for human rights and fundamental freedom; b. development of their personality, talent, and health to their fullest potential; c. should enable participation of the disabled in a free society. Article 24 also promises the assurance of quality and equal educational opportunities through inclusive education. It also mentions that to ensure facilities of educational institute for the upgrade the full and equal participation of disabled students, states parties shall ensure that disabled students can access i. General education and vocational education; ii. Adult education; and iii. Lifelong learning without discrimination (Convention on the Rights of Persons with Disabilities and Optional Protocol. 2021).

The main aim of SDG 4 is to ensure inclusive and equitable quality education for all that promotes lifelong opportunities. Regarding the education of children with disabilities, target 4.5 of SDG 4 focuses on providing the instruction that should not exclude anyone because of gender, safety, violence, disability, and education environment.

According to the SDG 4 reports, there are ten targets, and 12 out of 43 indicators are global. Globally, Target 4.5, inclusive education, was achieved by 35 percent of the world. According to available metadata, North Africa and Western Asia have 37 percent achievement, followed by Eastern and south-eastern Asia, Latin America, and the Caribbean, who all have 34 percent achievement (Framework of Sustainable Development Goals, 2021).

From the data, more progress is shown in the aspect of enrollment of disabled students in schooling, whereby Ethiopia has 98 percent and Papua New Guinea has 85 percent. India is in 8th position with 62 percent of enrollment, which shows that tremendous efforts are needed in India to promote the education of disabled children (Table 8.2).

8.2.1 SDG 4 in India

India is a country with states that ensure free education for all and any child who is below 14 years should get compulsory education. India is also like others who are setting indicators of achieves ambitious sustainable development goals. The targets for SDG 4 in India are presented in Table 8.3.

8.2.2 Policies and Programs

For the progress in achieving equality in education formulated main Acts especially for a child with disabilities.

Article 41 – Right of the disabled to get an education, work, and public assistance.

- Article 45 – Compulsory education act.
- Persons with Disability Act 1995 – Provides equal opportunities, protection, and full participation for the disabled.

TABLE 8.2 Enrollment of Children with Disabilities

Sl. No.	Countries	Percent
1.	Ethiopia (rural)	98
2.	Papua New Guinea	85
3.	Tanzania	85
4.	Indonesia	82
5.	Bangladesh	74
6.	Vietnam	69
7.	Malawi	68
8.	India	62
9.	Albania	54
10.	West Bank and Gaza	51
11.	Uganda	36
12.	Nigeria	28
13.	South Africa	21
14.	Maldives	15

Sustainable Development Goals, National Indicators Frame work Progress.

TABLE 8.3 Target Indicators for SDG 4 in India

Sl. No.	Target No.	Targets
1.	4.1	Ensure complete free, equitable, and quality primary and secondary schooling for all
2.	4.2	Ensure quality early childhood education
3.	4.3	Ensure affordable and quality technical, vocational, and tertiary education
4.	4.4	To increase the number of youth and adults with technical and/or vocational skills
5.	4.5	Inclusive education for vulnerable people with disabilities and also indigenous adults and children
6.	4.6	Achieve literacy and numeracy
7.	4.7	All should acquire the knowledge and skill to promote sustainable development
8.	4.a	Inclusive education for all children with disabilities and all the genders
9.	4.b	Increase enrollment in higher education by the increasing number of scholarships for developing countries
10.	4.c	Increase the number of qualified teachers and international coordination for teacher training

Source: SDG India Index and Dashboard, 2020.

- Rehabilitation Council of India Act – Monitors and regulates professionals and personnel, ensuring quality education through training.
- The National Trust – Ensures the welfare of persons with autism, cerebral palsy, mental retardation, and multiple disabilities to empower the disabled to achieve social position and financial empowerment.
- Action plan for Inclusive Education (IE) of children and youth with disabilities (2005) – Promotes IE for youth and children with disabilities in India (Taneja-Johansson et al., 2021).
- Right to Education Act (2009) – Education became a fundamental right according to the Indian constitution. RTE 2009 clearly states that no child should be failed or be expelled from education until elementary education completion. The right of persons with disabilities in education until 18 years comes under the persons with disability act, which came into force in 1995 to ensure equal educational opportunities for those with disabilities. This act provided good educational opportunities for the disabled, but learning for the disabled was not there (Taneja-Johansson et al., 2021).

The RPWD Act 2016 was enacted in December 2016 after India signed the UN Convention of the Rights of Persons with Disabilities (UNCRPD) Act. This act ensures the right to equality, life with dignity, and respect for their integrity equally remains in society for disabled persons. In the RPWD Act, education for the disabled is considered the duty of educational institutions of state government. These institutions should ensure the enrollment of students with disability, provide inclusive education, and that shall end (1) admit children with disabilities without discrimination and provide educational opportunities for all equally, (2) make infrastructure facilities accessible for the disabled, (3) offer sensible changes according to the individual needs, (4) provide support for the children with disabilities in inclusive education, (5) ensure appropriate language and means of communication facilities for children who are blind or deaf in inclusive education, (6) take measures for exact learning disabilities in children at the early stage, (7) evaluate and monitor the participation and progress of disabled students and (8) provide transport facility. The RPWD Act also has specific measures to promote and facilitate inclusive education by the government and the local authorities (Narayan and John, 2017).

Right to Education (RTE) – From 2009, education became a fundamental right in India. The RET act promises education for all children in the age group 6–14 years, and no child should be expelled and held back until they complete elementary education. The right to education for persons with disabilities until 18 years was ensured under the Persons with Disabilities Act 1995, which ensures educational opportunities for the children with disabilities (Kushwah, 2018).

Sarmagra Shiksha Abhiyan (SSA) – This is the most effective program for a zero-rejection policy. After the amendment to the constitution in 2002, compulsory education for children from 6–14 years and the result of the changes in schooling made conjunction with World Bank and made SSA in India. It is a program which concentrates more on the education of the specially abled rather than an inclusive education. Successful initiatives from SSA have improved the enrollment of students with disabilities in the education sector. This program allocates Rs. 1200 annually per child with disabilities, each district has the power to formulate plans to educate disabled students and train the teacher who handles those classes. There should be a pivotal institution to encourage collaboration to support children with disabilities (Bishnu, 2017). Block Resource Centres (BRCs) are established in each block of every Indian district and conduct in-service training. SSA came up with resource rooms to meet the needs of children with disabilities under inclusive education. These resource rooms are well equipped, which can help the children's therapy and education. SSA also provides training for teachers and parents: audio-visual aids, screening the child's needs and their performances, counseling, and exposures (NCERT, 2019).

Samagra Shiksha has three schemes which were launched in 2018: Sarva Shiksha Abhiyan (SSA), Rashtriya Madhyamik Shiksha Abhiyan (RMSA), and Teacher Education (TE) (https://vikaspedia.in/education/policies-and-schemes/samagra-shiksha, 2021). Sarva Shiksha Abhiyan is the program with a precise time amount for universal education design, quality primary education, and effective involvement of children with disabilities (Abhiyan et al., 2010).

New Education Policy – On 29 July 2020, national education policy was ratified by the union cabinet of India; it is set to transform the Indian education system by 2040. It aims to change the Indian education system into an equitable and vigorously knowledgeable society by providing quality education. In this policy, fair and inclusive education is also considered an excellent tool for achieving social justice and equality – this policy aligns with the RPWD Act and recommendations regarding the schooling of disabled children. An inclusive classroom should fulfill the needs according to the individual's disability. Quality educators and effective techniques are provided for a better education system for all by 2040.

There are many schemes for the empowerment of disabled students by the Ministry of Social Justice and Empowerment: Deendayal Disabled Rehabilitation Scheme (DDRS),

Assistance to Disabled Persons for Purchase/Fitting of Aids and Appliances (ADIP), Scheme for Implementation of Rights of Persons with Disabilities Act, 2016 (SIPDA), District Disability Rehabilitation Centres (DDRC), System of financial assistance for skill training of persons with disabilities, and Indian Sign Language Research and Training Centre (ISLRTC) (All India School Education Survey, 2016)

Schemes under the National Trust by Department of Empowerment of Persons with Disabilities, Ministry of Social Justice and Empowerment are Early Intervention and School Readiness Scheme, Vikaas (daycare), Sambhav (aids and assistive devices), and Badhte Kadam (awareness and community interaction). Schemes under the Department of School Education and Literacy (DSEL) Ministry of Education are: Samagra Shiksha Abhiyan – for the schooling of children with disabilities and Integrated Child Development Services for early childhood care and education. The scheme for early identification and intervention is Rashtriya Bal Swasthya Karyakram (RBSK).

Schemes for skill development and employment are: Vocational Rehabilitation Centres (VRC), Deen Dayal Upadhyaya Grameen kaushalya yojana (DDU-Gky) and Plan for Implementation of Rights of Persons with Disabilities Act, 2016 (SIPDA). The other schemes are Project of Sports & Games for the Disabled, Scheme of Assistance to National Sports Federations, Relief in Taxes to the family members of persons with disabilities.

Educational Status of Children with Disabilities

The target 4.5 is a global indicator that ensures the education for previously excluded groups in society by 2030. Children with disabilities have been neglected in education. As specified by the metadata provided in UNICEF's SDG reports, only 19.3 percent of children with disabilities who are 15 years and above have completed at least secondary education in India. No state or union territories have achieved the target of 100 percent, which is scored by measuring their performance in SDG 4 targets. The best performing states in India are Goa with 32.4 percent, and (among Union Territory) Delhi with 41 percent. In Arunachal Pradesh, only 10.3 percent of persons with disabilities have completed secondary education (SDG 4 Data Digest SDG 4 Data Digest, Indicators, T. E. 2019).

From SDG report 2021, metadata shown in Table 8.4 is about the front-runner according to their target 4.5 and the total index score on SDG 4 in India. According to the NITI Aayog report, India is a performer, even though no state or UT became an achiever of the

TABLE 8.4 Runners Up of Target 4.5 in India 2020–21

Sl. No.	States and Union Territories	Target 4.5	Index Score
1.	Kerala	24.3	80
2.	Chandigarh	37.4	79
3.	Delhi	41	75
4.	Himachal Pradesh	25.6	74
5.	Goa	32.4	71
6.	Puducherry	28.6	70
7.	Uttarakhand	24.7	70
8.	Tamil Nadu	19.1	69
9.	India	19.3	57

Source: SDG India Index and Dashboard, 2020.

SDG 4. According to the index score, Kerala is the front-runner, but Delhi has the highest rate in providing education for previously excluded groups in society (target 4.5).

8.2.3 Enrollment of Children with Disabilities

According to the 8th All India School Education Survey, only 21.11 percent of schools are standing by the commitment to inclusive education for disabled children. The proportion of schools adhering to inclusive classrooms in India are: primary schools, 60.4 percent; upper primary schools, 28.33 percent; secondary schools, 6.59 percent; and higher secondary education schools, 5.07 percent. The proportion of teachers who have received training is 1.32 percent; the number of special educators who never visit the schools is 42.69 percent; and frequent visitor teachers for specially-abled students is 57.31 percent. The rate of differently abled students enrolled in primary school is 52.27 percent; upper primary school is 28.78 percent; secondary is 15.5 percent; and higher secondary school is 3.45 percent. According to the enrollment data, 29.16 percent of disables students have a visual impairment; 14.47 percent have a hearing impairment; 25.05 percent have an orthopedic disability; 22.35 percent have an intellectual impairment; 4.20 percent have multiple disabilities; and 4.77 percent have other types of disabilities (All India School Education Survey, 2016).

According to data, most of the children with disabilities completed upper primary education compared to other education levels, and the children with disabilities in higher secondary education ratio is lower than the enrollment ratio of upper primary schooling (Figure 8.1). Efforts need to be made to expand the enrollment rates of children with disabilities in higher education too.

8.2.4 Infrastructural Facilities

Providing infrastructural and educational facilities attract students toward education. But because not all schools are fully equipped, the distance prohibits the excluded groups in society from getting to suitable facilities. Target 4.a is focused on building and upgrading

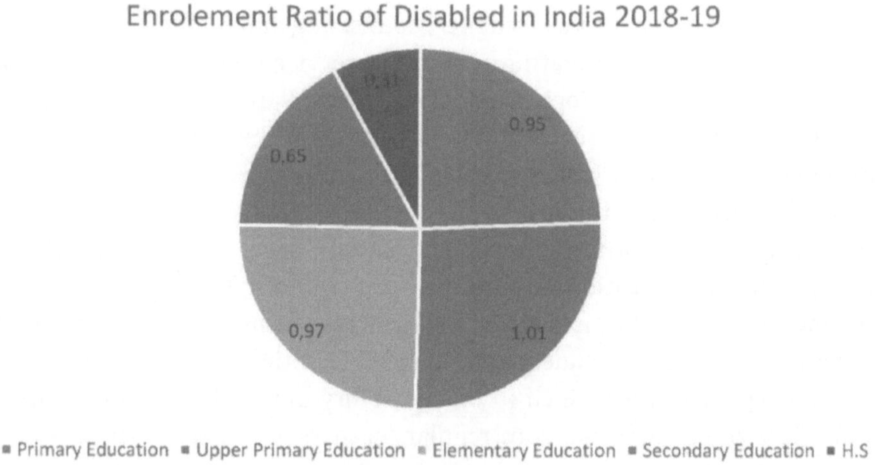

FIGURE 8.1 Enrollment ratios of children with disabilities in India (2018–19). (Source: SDG India Index and Dashboard, 2020).

infrastructural facilities in schools for disabled and other excluded groups in India, so that these initiatives can change the situation for the better. There are eight front-runner states in the achievement of SDG 4: among them Delhi and Chandigarh are achievers in target 4.a; followed by Goa with 99.97 percent Pondicherry with 99.93 percent; and Kerala with 99.24 percent. Himachal Pradesh, Tamil Nadu, and Uttarakhand are also on the way to providing infrastructure for the disabled and students from vulnerable situations in their districts (SDG India Index and Dashboard, 2020).

Among all schools in our country, only 10 percent have handrails, 43.62 percent have ramps, 3.81 percent have adapted laboratories, and 7.64 percent have adapted lavatories (All India School Education Survey, 2016). The proposition of major infrastructures like toilets and ramps, which are friendly to the disabled, are 17.54 percent and 63.71 percent, respectively, in inclusive schools in India. Information and communication technology (ICT) and computer labs for quality education and better adaptation for the disabled are only 32.66 percent, which should improve more to make education exciting and attractive for the students.

8.2.5 Resources, Teachers, and Inclusive Education

According to a National Informatics Centre (NIC) report, 3500 teachers and heads of schools completed in providing facilities and expert training sections. Teachers in inclusive schools have networked with teachers and health officers to provide better education for students (NIC Report on Inclusive Education, 2019–20). Teachers in inclusive education are tasked with identifying children with disabilities, referring students for expert examination and treatment, including and accepting disabled students in the classroom, placing children with disabilities according to their needs at the school and making them feel comfortable, ensuring that infrastructural barriers are changed or removed to allow independent movement, involving them in classroom activities and educational discussions, making suitable adaptations in the curriculum so that children with disabilities can learn according to their ability, using teaching aids that fulfill the needs of children with disabilities, providing parental guidance, counseling and public awareness programs through school activities; collaborating with other service sectors for children with disabilities, like medical and physiological, and social work. Parents and exceptional teachers should construct unique rating scales and tools to analyze the achievements in academics and improvement in disability, adapt suitable evaluation tools for children with disabilities, give remedial advice to children who need it (Tyagi. G, 2016).

Resource teachers who work in government schools face social/personal challenges in implementing inclusive education in rural primary schools. The appropriate number of teachers in inclusive education is deficient, so general teachers handle the class without accurate technical knowledge. Inadequate transport facilities in India, a minimal number of resource teachers, teachers trained in one disability must take care of all the disabled children, and lack of cooperation from regular teachers are some of the challenges that resource teachers faced in implementing inclusive education. Schools without proper infrastructure facilities and teaching–learning materials also add to their burden (Umapathi and Maithily, 2020).

8.3 NITI AAYOG STRATEGIES

NITI Aayog is an SDG measuring organization which provides data on achievements. It is evolving in sharing visions of the nation in development in prior sectors and setting up strategies with the effective involvement of states in the light of national objectives. It also develops village-level plans and makes them progress to the higher government. NITI Aayog provides advice and encourages partnerships between stakeholders and government for the development of educational and policy research institutions. According to these functions, derived strategies of NITI Aayog for new Indian education are:

- To change society's attitude toward the disabled, including more case studies and success stories of persons with disabilities in the ordinal curriculum.
- Along with teacher's regular training sessions, include special education training also.
- Scholarships and fellowships for persons with disabilities should be enhanced.
- To improve inclusive education in India, schools should address their environmental barriers, admission procedures, and curriculum design.
- Ensure the schools have a section of each special classroom available for disabled students with others is comes under Universal Design for Learning (UDL) guidelines.
- Promote collaborations among inclusive and special schools in government and private sectors, fostering partnerships with MHRD and MSJE.
- Develop rating scales for measuring the school's activities toward inclusive classrooms.
- Inclusive infrastructure in schools should also provide opportunities in sports and cultural activities for disabled students.
- The government will spend more on education as a whole up to at least 6 percent of GDP by 2022 (N for Nose, 2019).

8.4 CHALLENGES IN INCLUSIVE EDUCATION

Inclusive education is a newly implemented educational strategy that needs to improve more to achieve its goal. Even finding suitable teaching technologies for these classrooms and their needs are found by implementing new actions. There are numerous challenges and barriers to clearing the path for inclusive education in India. Mainly the challenges are discussed in two aspects of inclusive education.

Challenges in the schooling of persons with disabilities in India are:

- Lack of qualified and trained teachers: According to the surveys, the lack of skilled and professional teachers for educating the disabled in inclusive classrooms is the biggest challenge in the current situation. There are more than one million vacancies for general teachers (NIEPA, 2014). Only 210,646 special education teachers are appointed under the scheme of SSA across the country (i.e., nearly two-thirds of the vacancies have not been filled yet (Singh, 2016).

- Attitude and capacity of teacher handling inclusive classrooms: teacher should need an adaptive attitude for the children with different needs in a school. There are many training programs for teachers, but the overall quality and effectiveness of the programs are a concern for the government (Government of India, 2015, David P, 2018) (.
- Teacher's curriculum: inclusive education courses have been recently added to the teacher's education curriculum due to a lack of awareness of the concept of inclusive education. It is necessary to engage with the students in an inclusive classroom to teach the concept well (David. P.,2018, Singh, 2016)
- Instructors of teacher's training: we are in the initial stages of teaching inclusive education, where there are not so many experienced trainers in this field (Singal, 2005).
- The ability of special educators: special educators need more training to handle a child with a single disability. There are about 27 disabilities recognized in the RPWD Act in India. Educators' lack of skill and experience in this field is a challenge. The model followed in India is a special educator handling a range of disabilities through multiple categories, which is unrealistic and criticized by the experts (Allan, 2014).
- Academic support structure: SSA created a plan to support educators at the district and subdistrict levels for various reasons (GOI, 2011). The students and the teachers face problems in a diverse classroom due to a lack of support from administration, non-availability of assistive devices, special educators, and others (Bika & Education, 2021).

Some systemic issues are also challenges in inclusive classrooms which include the following:

- Lack of on-site resources: the number of special educators in inclusive education is quite small compared to the need, so school visits are infrequent. Parents are encouraged to take their children with disabilities to the Cluster Regional Centres (CRCs) or Block Resource Centres (BRCs) for weekly therapy and skill development. Some states don't have CRCs where parents could take their children to BRCs weekly once. As a result, parents are unable to travel far for weekly resource utilization Allan, 2014).
- Experience of parents: adequate transport and effort of interaction with the institution are highlighted problems by parents of disabled children in India. Some parents also spend time in schools looking after their children; if they have other work, they should bribe guards or non-teaching staff to keep an eye on their children. Some parents travel long distances daily to drop off and pick up their children (David P., 2018)
- The expectations of parents and teachers toward the academic performance of disabled students: typically, they have low expectations for children with disabilities on their academic performance. The parents always want their children to be socially

active and improve their vocational skills; they are not concerned about learning ability (David P., 2018).

- Labeling: one significant issue is labeling children with disabilities with the appropriate label of their condition. Sometimes choosing proper treatment and teaching techniques is not enough (Bishop, 2014).
- Lack of research: little research has been conducted about the resource teacher's efforts and needs in inclusive education (David P., 2018).
- Large society and highly competitive nature of education: the community's lack of involvement and awareness toward inclusive education is a major challenge affecting the enrollment of disabled children in schooling. Most of the barriers experienced by disabled pupils are from the ways of thinking among the community. A highly competitive education system is a problem not only for regular students but also for the disabled (State of the Education Report for India, Children with Disabilities Report, 2019).

The significant challenges in inclusive education are listed in the AISES 8th survey:

- Among all the schools in the country, only 21 percent are adapted for inclusive education for disabled children (All India School Education Survey, 2016).
- Out of these schools, the proportion of primary, upper primary, secondary, and higher secondary schools are 60 percent, 28 percent, 7 percent, and 5 percent, respectively, in India (All India School Education Survey, 2016) (Mann, 2021).
- Teachers who have received training to educate the disabled represent 1 percent out of all the teachers.

  (https://friendsforinclusion.org/2019/12/27/findings-for-people-with-disabilities-from-the-eighth-all-india-school-education-survey/).
- Braille books are available in only 7 percent of the schools that have students with visual impairments.

  (https://friendsforinclusion.org/2019/12/27/findings-for-people-with-disabilities-from-the-eighth-all-india-school-education-survey/).

8.5 BARRIERS TO ACHIEVING SDG 4

8.5.1 Globally

Barriers to educating excluded groups in society include the following three factors:

- Policy and system factors: most policies do not encourage students with disabilities to attend school, and the procedures are dated and inappropriate or based on the medical approach to children with disabilities. There are reasonable policies that are not implemented well and poor resource allocation for inclusive education. There are limited training and a lack of incentives for teachers who handle disabled students. Poor school structures and supporting services in many countries make inclusive education a challenge (Guidance note: A DFID practice paper, 2021).

- Social and communication factors: the most significant barrier to inclusive education is due to society, not the disabling impairment, according to UNESCO. The attitude of parents and social stigma toward the disabled result in society considering those with disabilities as being 'punished'. Typical barriers like financial and transportation (Guidance note: A DFID practice paper, 2021).
- School factors: budget of school administration for appropriate needs of students; inadequate training for the teacher to handle inclusive classrooms; lack of awareness about disability among teachers and staff (Guidance note: A DFID practice paper, 2021).

8.5.2 India

The major problem that India is facing for achieving SDG goals is financing the goals according to the efficiency of states. To make equal opportunities for all in education will take filling the gap of infrastructural and technical needs. Monitoring and evaluating is also a big deal for India; NITI Aayog is the only organization that tracks progress. Measuring the progress is also a challenge because India's progress in SDG has not been entirely transparent. So a separate measuring index for the progress evaluation and monitoring is also needed (Sustainable Development Goals-Challenges for India, 2018). Policy is always written in the top-down method; that's why the problems in inclusive education are not addressed during policy making. Also, educators and schools are not allowed to be involved in policy and program-making; this is another barrier to the success of inclusive education (Hupe et al., 2014).

Due to Covid-19, the whole world seems to have fallen back in becoming an achiever in the performance of SDGs, especially SDG 4. This pandemic period teaches us how much technical support is needed to fill the gaps. Assistive technology tools in education should include SDG 4 targets for overcoming the barriers in education (Table 8.5).

8.6 SUGGESTIONS

8.6.1 Global Level

- A broad range of legislative and policy measures is needed. It's not sufficient to introduce inclusive education through programs and policies but also needs implementation of a strategic approach and commitment.
- Frame indicators should be adopted by all the countries, developed and developing alike.
- Assistive technologies and support to improve technical skills of teachers to deal in an inclusive classroom should be included.

TABLE 8.5 Schools with Infrastructural Facilities for the Disabled in India 2018–19

Sl. No	Facilities	Percent
1.	Electricity	73.86
2.	Computer labs/ ICT classrooms	32.66
3.	Adaptive infrastructure	
	I. Toilets	17.54
	II. Ramps	63.71
4.	Sanitation facilities	
	I. Boys	92.63
	II. Girls	94.38
5.	Water facilities	
	I. Drinking water	95.68
	II. Hand wash (as per WASH indicators definitions)	88.18

Source: SDG India Index and Dashboard. 2020.

8.6.2 National Level

- Ensure sufficient resource teachers are available to fulfill the need of students with disabilities.
- Construct appropriate teaching approaches and assistive technologies suited for the inclusive classroom.
- Should encourage the academics of students as well as the social interaction and emotional stability.
- Should implement separate index measuring tools and update with the achievement level of each state. Through this data, the efficiency of implemented policies can be analyzed.
- Effective study materials that can improve the academic performance of children with disabilities must be provided.

8.6.3 State Level

- Design effective classroom plans and strategies.
- Construct teaching strategies in the inclusive classroom according to the Universal Design in Education principles.
- Provide appropriate training for teachers and parents to educate students with disabilities.
- Implement study material and teaching methods or tools which can improve the academic performance of disabled students in inclusive education.

8.7 CONCLUSION

Functional changes that transform ordinary classrooms to inclusive classrooms are happening because of SDG 4. Quality education for disabled students became the 'hot topic' of discussion in many countries after SDG 4 inclusive classrooms provide advanced teaching–learning experiences for all the students – even the 'abled' ones. Taboos and negative attitudes toward disabled students are changing; the future education system will make drastic changes in society. Through these, there will be more higher education and job opportunities for the disabled. The country's complete development will be when women, the disabled, and youngsters are economically and technically empowered. Achieving SDGs in each country is challenging, even in light of the tremendous human force to implement them. Schooling for the disabled, vulnerable, and gender-sensitive can create significant economic stability for the future of our country.

REFERENCES

Abhiyan, S. S., Sarva, I., Abhiyan, S., & Cabinet, U. (2010). SSA. Education, E., Performance report, Ministry of Human Resource Development. 15, 1–51.

All India School Education Survey. (2016). Education Survey Division. National council of Education research and training. ISBN 978-93-5007-764-1

Allan, J. (2014). Inclusive education and the arts. Cambridge Journal of Education, 44(4), 511–523. https://doi.org/10.1080/0305764X.2014.921282

Bika, S. L., & Education, D. (2021). Bring the Child with autism towards inclusive education: A case study. *Levant Journal.* 20(7), 110–117. ISSN: 1756-3801.

Bishnu, S. (2017). Policies for developing inclusive education system: A review. 3, 3312–3316.

Bishop, D. V. M. (2014). Ten questions about terminology for children with unexplained language problems. International Journal of Language and Communication Disorders, 49(4), 381–415. https://doi.org/10.1111/1460-6984.12101.

Convention on the Rights of Persons with Disabilities and Optional Protocol (2020) United Nation. pp. 1–5.

David, P. (2018). Sustainable Development Goals Challenges in India. *Indian Journal of Public Health Research and Development* 9(3), 1. DOI:10.5958/0976-5506.2018.00172.9.

Framework, N. I. (2021). Sustainable Development Goals. United nation. https://unstats.un.org/sdgs/report/2021/The-Sustainable-Development-Goals-Report-2021

Guidance Note: A DFID Practice Paper. One, S., & Information, B. (2021). Education for children with disabilities: Improving access and quality. 1–22.

GOI, (2011). https://darpg.gov.in/sites/default/files/Sarva%20Siksha%20Abhiyan

GOI, (2014). SSA. https://accountabilityindia.in/publication/sarva-shiksha-abhiyan-goi-2014-15/

Hupe, P., Hill, M., & Nangia, M. (2014). Studying implementation beyond deficit analysis: The top-down view reconsidered. Public Policy and Administration, 29(2), 145–163. https://doi.org/10.1177/0952076713517520

Kushwah, I. B. (2018). International Journal of Advance Research, Ideas and Innovations in Technology. Right to Education and Learning Disabled Children. 4(4), 701–703. www.IJARIIT.com

Mann, B. (2021). Report on the Condition of Education 2021. P 34. https://nces.ed.gov/pubs2021/2021144.pdf

Narayan, L. C., & John, T. (2017) The Rights of Persons with Disabilities Act, 2016: Does it address the needs of the persons with mental illness and their families. Indian Journal of Psychiatry, 59(1), 17–20 doi: 10.4103/psychiatry.IndianJPsychiatry_75_17.

National Sample Survey Office. (2016). Disabled persons in India: A statistical profile. 0–107. http://mospi.nic.in/sites/default/files/publication_reports/Disabled_persons_in_India_2016.pdf

NCERT (2019). Chemistry: Textbook for class XI - Part. I. National Council of Educational Research and Training.

NIC report on inclusive education, 2019–20 http://www.edudel.nic.in/samagrashiksha/content/2020/inclusiveeducation.pdf

NIEPA (2014) Annual report.

N for Nose (2019). State of the Education Report for India 2019 Children with Disabilities. UNESCO, p. 84. ISBN. 978-81-89218-67-6.

SDG India Index and Dashboard (2020). NITI Aayog. Govertment of India. pp. 93–101.

SDG 4 Data Book, Global Education Indicators. UNESCO (2019). UNESCO institute for statistics. pp 35– 63. ISBN 978-92-9189-235-8.

SDG 4 Data Digest SDG 4 Data Digest, Indicators, T. E. UNESCO (2020). UNESCO institute for statistics. ISBN. 978-92-9189-252-5.

Sustainable Development Goals, National Indicators Frame work Progress Report. (2021). Ministry of Statistics and Programme Implementation. www. Mospi.

Singal, N. (2005). Mapping the field of inclusive education: A review of the Indian literature. International Journal of Inclusive Education, 9(4), 331–350. https://doi.org/10.1080/13603110500138277

Singh, D. (2016). Inclusive education in India: Concept, need and challenges. Scholarly Research Journal for Humanity Science and English Language, 3(January 2016), 3222–3232.

Taneja-Johansson, S., Singal, N., & Samson, M. (2021). Education of children with disabilities in rural Indian government schools: A long road to inclusion. International Journal of Disability, Development and Education, pp 1–16. https://doi.org/10.1080/1034912X.2021.1917525

Tyagi, G. (2016). International Journal of Education and applied research. ISSN: 2348-0033. 06 (01).

Umapathi, N., Pushpa, K. S. and Maithily, K. S. (2020). Implementation of inclusive education-resource teacher faced by challenges at primary school. Juni Khyat(UGC Care Group I Listed Journal), 10(9), 39–41. ISSN: 2278-4632

UNESCO. (2019). Discussion on SDG 4: Quality education. High-Level Political Forum on Sustainable Development, July, 14. https://sustainabledevelopment.un.org/content/documents/23669BN_SDG4.pdf

United Nations. (1994). Salamanca-Statement-and-Framework.in Action.

United Nations. (2016). Sustainable Development Goals Repot. https://www.un.org/development/desa/publications/sustainable-development-goals-report-2016.html.

WHO. (2011). Summary world report on disability. World Health, 1–24. www.who.int/about/licensing/copyright_form/en/index.html%0Ahttp://www.larchetoronto.org/wordpress/wp-content/uploads/2012/01/launch-of-World-Report-on-Disability-Jan-27-121.pdf

CHAPTER 9

Techno Pedagogical Concept with Digital Skills

Sugandha Agarwal, Deepa Gupta, and Mukul Gupta

CONTENTS

9.1 DEFINING LEARNING

In a different context, learning is defined differently. According to Stephen P. Robbins 'learning is the modification in behavior due to differential experiences' (Saravanakumar, 2019). Learning includes two important variables, i.e., there should be a permanent change in behavior and this change occurs due to continuous practice (Tomar, 2014). For example, driving a car: here, ability to drive is the permanent change, and with constant practice,

DOI: 10.1201/9781003254942-9

this expertise was possible. To bring changes in anyone's behavior, it is required that there must be a certain purpose associated with it. Without goals and targets, there is no motivation for anyone for learning. For example, becoming independent may act as motivation for a housewife to learn driving. An urge to reduce traveling time and dependency on others acts as a cue for the housewife to learn how to drive. In response to which she will enroll herself in driving school and get training on driving skills. When the driving skills are acquired by constant practice, the difference in smooth operation is easily observed, and hence learning is achieved. So, the learning process includes motivation, cues, response, and reinforcement (https://courses.lumenlearning.com/). Any kind of learning comprises awareness, intellectual skills, insolence, emotional response, and communal behavior (Mehmet Kesim, 2015). According to the educational definition, learning is the progression of people obtaining talents, knowledge, empathy, and other qualities (Robinson, 2012).

9.2 TYPES OF LEARNING

Fredric Vestor identified four types of learning in his book Denken Lernen Vergressen, published in 1975.

1. Learning through the eyes is called 'visual learning'.
2. Learning by listening and speaking is called 'auditive learning'.
3. Learning by touching and feeling is called 'haptic learning'.
4. Learning through the intellect is termed 'reading/writing' learning (see Figure 9.1).

FIGURE 9.1 There are four types of learning. 1) Visual 2) auditory 3) haptic 4) read /write.

Visual learning: anything which we see remains for a longer period of time and when the observed process is repeated again and again, the learning becomes permanent (Raiyn, 2016). For example, the very first skill which an infant learns is speaking which he learns by observing the movement of the mouth of his parents and imitating the movement. Watching movies, documentaries, videos, etc. creates a larger effect on our brains than only listening.

Auditory learning: here the individual learns when there is some reinforcement of sounds. Learning occurs after listening to instructions and following them (Kayalar, 2017). For example, along with observing the movement of mouths, when an infant listens to the sounds associated with it, then he can imitate and speak properly.

Haptic Learning: in this type of learning, the individual tries to touch and handle the activity been done so that he/she can experience it and learn to perform the same activity properly. Dance, sports, swimming, etc. are activities where one can learn the skill by performing it by oneself (Mitchell, 2018).

Reading/writing learning: this type of learning includes intellect, where the individual learns through written words. It is also termed 'classroom learning' as our traditional learning system is based on writing, reading books and articles, taking notes, looking up the relevant material required to gain knowledge.

9.3 LEARNING THEORIES

(See Figure 9.2).

1. Cognitive Learning Theory: in cognitive learning, the maximum brain potential, effectiveness, and application are used, in which the person's ability to contemplate the event, perceive the event, evoke the event, and explain the problem is applied (Enliang Yan, 2021). This is the initial way of learning used from the time of our childhood; a child learns all the activities and achieves their milestones through cognitive learning (Enliang Yan, 2021). The use of cognitive learning if well applied in the education or professional environment is an exceptional way to nurture learning in the organization (John Dunlosky, 2013). It helps to build confidence, increase the pace of learning, encourage inclusive learning, enrich problem-solving skills, and 'kick-start' innovation.

2. Behaviorism Learning Theory: when an individual learns due to certain interactions with the environment, and a systematic and observable study can be conducted in which stimulus–response behavior is generated due to some specific conditioning, this process is called behaviorism learning theory (Guillaume Fontaine, 2021). The classical conditioning experiment of Pavlov through positive reinforcement of dogs proved the theory (Baker-Rush and Pabst, 2021). In professional life, incentives for hard work always act as positive reinforcement for employees to perform better.

3. Constructivism Learning Theory: in this, students construct the knowledge instead of just taking in information. Whatever information is received by the student is then connected to the previous learning and creates their learning (Sitthiphong

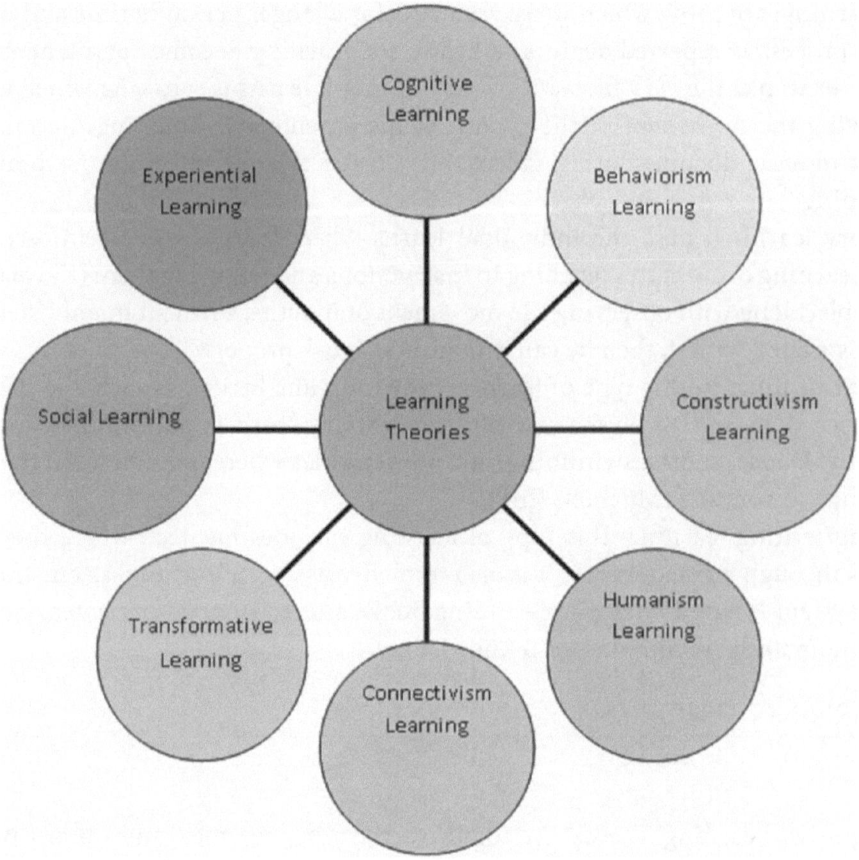

FIGURE 9.2 Learning theories.

Suwannaphisit, 2021). This learning process is active, personal, and individual for each student. The role of a teacher in this kind of learning is more of a guide, or mentor, so that student can build their own understanding and wisdom (Jan Moons, 2013).

4. Humanism Learning Theory: this type of learning is based on the five levels of Maslov's Hierarchy of Needs theory. When the first four needs of the student or individual are met, he/she strives for the final need, that is, self-actualization (Sanmugarevathi, 2020). If the guide/mentor/teacher provides a conducive environment, then a person can learn to attain self-actualization or the means to attain it (Madsen, 2012).

5. Connectivism Learning Theory: this is the modern approach to learning and gaining popularity. According to this theory, it is believed that a student or individual learns more quickly when they are/become connected to the concept or event. Hobbies or any activity which we do practically, or like to perform, or anything which we want to hear and discuss, gain the quick attention of the student and enhance learning (Shrivastava, 2018). Today trainer/teacher role is more about creating a learning environment, developing the connection with the concept so that students can attentively grasp the information, and learning occurs.

6. Transformative Learning Theory: this type of learning is more applicable to adults, where they gain a new perspective and apply the same in the dynamic environment (Yun-Jeong Oh, 2021). For students, they must question and have an open discussion clearing their all doubt/myths related to any perspective they had learned or observed from childhood. This learning focuses on evaluating past ideas, beliefs, and experiences and then critically analyzing them with the new information and environment so that a new understanding of what to believe and follow can be developed.
7. Social Learning Theory: this theory is based on four basic elements, i.e., attention, retention, reproduction, and motivation (Proctor and Neimeyer, 2020). Attention relates to conducting the activities to support student concentration. Retention is observing how the learner will adopt information and evoke it in the future. Reproduction is the depiction of beforehand erudite behavior and when it's suitable to apply it. Motivation is getting influenced by watching other getting rewarded or punished for their actions.
8. Experiential Learning Theory: when a student or individual learns by gaining experience of an activity, it is retained for a longer period. It includes elements like concrete wisdom, thoughtful scrutiny, grasping experience, conceptualization, and vigorous experimentation. An environment of learning and experience is provided so that whatever is been taught to the student can be practiced and get an insight into the real-world experience (Boyan Bontchev, 2018).

Theories of learning have come a long way and history has witnessed the changes in the ways knowledge is imparted and learning is performed. Learning also differs due to the age of the student and the context of learning. With the changes in the ways learning is grasped, it becomes essential that the way learning is imparted should also change (Kibble, 2016). Knowledge, education, wisdom, etc. are not new words that were generated just 100 or 200 years back; they have been in use from the time of our Hindu Vedas. According to Hindu mythology, there were maha gyani, maha pandit having immense knowledge. They had gained knowledge from some of their gurus. These gurus used to apply all the above-mentioned ways to impart knowledge to their shishyas. So, a Teacher/Trainer/Mentor, would focus on how to impart knowledge and enhance the learning of their Student/Trainee/Mentee.

9.4 TEACHING – FROM GURUKUL TO SCHOOL

9.4.1 Gurukul

As the name defines, 'gurukul' is the home of the guru (teacher), where shishya (students) were supposed to live and gain knowledge from their gurus. It is the oldest form of schooling rooted in the Indian subcontinent way back to around 5000 BC (Soni, 2018). It exists from the Vedic time, where highly focused education was imparted with equal treatment regardless of social difference. It created a tradition of guru–shishya where students were distributed into three categories: Vasu – students obtaining education up to the age of 24; Rudra – students obtaining education up to the age of 36; and Aaditya – students obtaining education up to the age of 48 (Soni and Trivedi, 2018). As with modern schooling, there were no fees involved in the gurukul system. Shishya was supposed to do all the personal chores by themselves and live a simple and satvik life. This methodology motivated students to develop discipline, patience, respect, and good habits. They were supposed to live in natural environments with humanity, fraternity, and love among themselves. The majority of the teaching was based on group discussion, self-learning, etc. on various subjects like language, science, and mathematics. Apart from these regular subjects they were involved in arts, sports, crafts, singing, meditation, yoga, mantra chanting, etc. to have positivity in the mind and to make them physically and mentally healthy. All these all activities helped the student navigate the real world as they cultivate the persona, self-reliance, brainpower, and sense of discipline in life. Gurukul education structure is a pyramid of intelligence, skills, and principles that revolves around knowledge (Selvamani, 2019) (see Figure 9.3).

Pupils in gurukul were required to recite the Vedas, as taught by the gurus, and gain hands-on experience, which benefited to gain strong memory power (Luniya, 2012). It was the most sacred and respected system of education. There was no fixed fees system involved, but it included the guru dakshina (honorarium), which needs to be given by the shishya, not the parents, as it is in the current education system. This guru dakshina can be in any form, like money or a task to be performed by the shishyas (Saranya Kumar, 2016). This can be understood by the story of Eklayva from Mahabharat where Eklavya had given his thumb of the right hand as a guru dakshina to guru Dron. Although it is criticized that the act of guru Dron was unjust and has been done just to prove Arjun as the greatest archer. But the real reason behind his act was the commitment of guru Dron to the kingdom of Hastinapur. If Eklavya would have been able to use his archer skills, then he may have used them against the kingdom of Hastinapur. So, at that time, no one questioned the acts of gurus, as it was considered that these gurus had the supreme knowledge of dharma.

9.4.2 Modern Education System

In the year 1835, the modern education system was introduced to India by Lord Macauley. It has evolved with time and is majorly influenced by western culture. Unlike gurukul, where students were supposed to live in the house of the guru, there are institutions like schools, colleges, and universities where the formal method of education is imparted. As this system was introduced during the British era, the major emphasis was to educate the

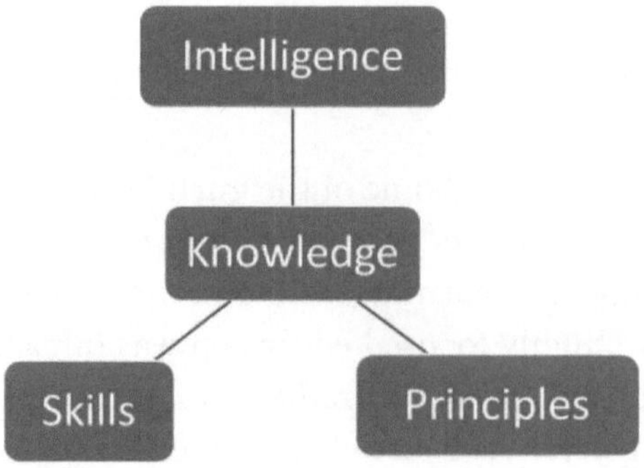

FIGURE 9.3 Gurukul Education Pyramid.

people in urban areas so that they could be employed in government offices under the rule of the British. The medium used to impact education during British time was English, and the Indian language was ignored. A University commission was established in 1902 with two Indian members, and it issued regulations and policy in 1904 (Mukherji, 1964). Numerous activities related to educational development were carried out after 1921. After 1929 and due to the Hartog Committee, the three levels of education, i.e., primary, secondary and higher education, were established (Chutia, 2020). The introduction of vocational and industrial-based education with linkage to employment was also suggested (Chaliha and Borah, 2016). Diversification in curriculum according to regional demand was also suggested: for example, agriculture and farming practice was essential in the rural areas, so subjects were developed in such a manner to promote agriculture in rural areas. In the same way, other schools were formed which focused on production, manufacturing, handicrafts, craftsman, artwork, and artisans. The goal of education was to augment the intellectual, moral, and socioeconomic status of the people. The pedagogy system and syllabus were designed in such a manner to deliver knowledge for the benefit of the individual and promote the development and growth of the nation (Mookerjee).

9.5 DIFFERENCES BETWEEN GURUKUL AND THE MODERN EDUCATION SYSTEM

The education system is interested in the acquisition of the two major skill types, i.e., hard skills and soft skills. Hard skills focus on learning subjects like literature, history, geography, social studies, civics, arts, and science. Whereas soft skills focus on mind development communication skills, intelligence quotient, emotional quotient, interpersonal relationships, team-building, and moral values. In the gurukul system, learning starts with the soft skills and moves toward the hard skills with age. A child was first taught the mind skills, aptitude, and soft skills, and less emphasis is given to brain training, technical data, and hard skills. As the child grows and is ready to become an adult, the focus of his education turned toward hard skills, brain training, and technical data. According to Hindu mythology, it is believed that building moral values, inculcating good habits, respect for others, discipline, self-awareness, understanding, and handling emotions, are developed during childhood, stays for life long, and build strong character. The hard and technical skills can be grasped at the adult stage, when the student is more focused and understands the direct benefit of learning the skills (Karve, 2013). Here the major difference between gurukul education and modern education is the way skills are imparted. According to gurukul system, first mana, which means the mind, is developed by building a receptacle and developing soft skills. Later technical training is imparted. Whereas modern systems first focus on budhi, that is brain development, by teaching hard and technical skills and later focuses on soft skills, IQ, EQ, and interpersonal skills (Frederick, 2016) (see Figure 9.4).

The major difference between the modern schooling and gurukul system are:

1) In gurukul, students were supposed to live at the guru's house, whereas in the modern education system, specific institution infrastructure is developed, and qualified teachers are hired to educate the students.

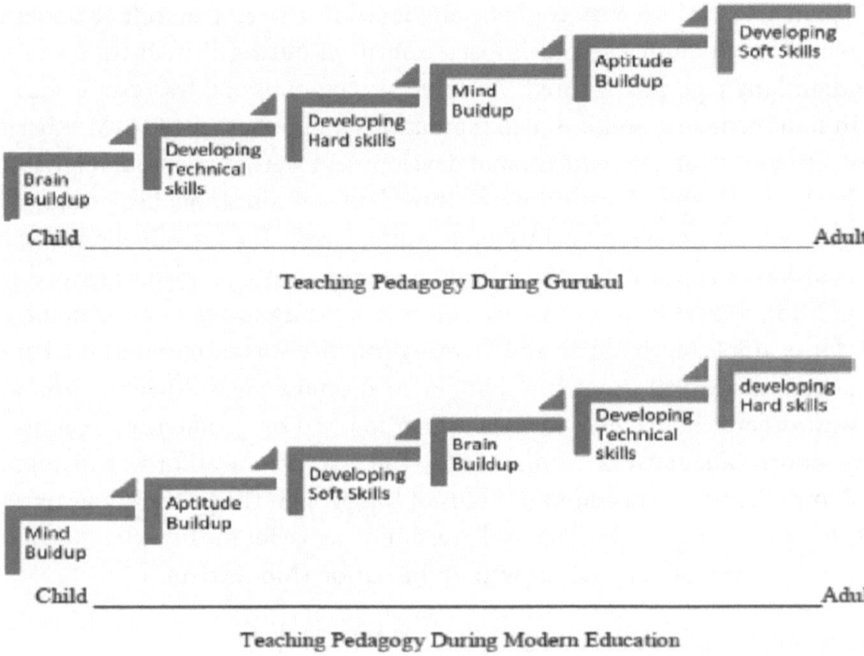

FIGURE 9.4 Gurukul training methodology.

2) Admission age in gurukul was around 8–12 years, whereas in modern schooling it is 3–5 years.

3) To gain admission to gurukul, one has to undergo the sacred thread ceremony called 'upanayan', whereas in modern schooling a basic exam needs to be cleared by the student.

4) Women's education was not appreciated in the ancient gurukul system, whereas men and women both can have formal education in the modern schooling system.

5) The method of rote learning, i.e., oral learning, was followed in the gurukul system, whereas today books, notes, pictures, videos, etc. are the way of imparting knowledge.

6) In gurukul system, the syllabus includes the discipline from the four Vedas were *purans, ithihas*, grammar, logic, ethical knowledge, astrology, maths, and *yudhshastra*, whereas modern schooling education leans more toward the theoretical knowledge of various areas.

7) In gurukul system, there were no formal fees charged, but after the completion of the course, shishyas were supposed to give guru dakshina which can be anything requested by the gurus. It may be in monetary terms or in any other form such as a task. In the modern education system, a fixed fee structure is designed and to be paid from time to time, the burden of payment in this system lies with the parent or guardian of the student.

8) In gurukul system, there are already established and renowned gurus under whom learning occurs, whereas in the modern education system qualified teachers are hired by the school's owners which may be state-owned or privately owned schools.

9) In gurukul system, all the learning takes place under one single guru so there is a lack of dynamism in the mode of study, whereas in modern teaching there are separate teachers for the different subjects.

10) In gurukul system, there is no set curriculum to be followed and there is no one to keep a check whether everything has been taught by the gurus. It relied totally on the trust and worship of gurus. In the modern system, there is a designed curriculum to be followed by the teachers, and it is monitored by the senior officials.

9.6 TEACHING PEDAGOGY

Teaching implies imparting education, knowledge, and skills to the learner. It includes all delivery of the concepts and information by the teacher to the student. Pedagogy defines the method to be followed for the delivery of education to the student (Cuenca, 2010). Teaching pedagogy encompasses teaching styles, theory, feedback, and assessment. Based on teaching experience, knowledge, and context a teacher develops the methodology to teach the concept in the class (Navaneedhan, 2011). There will be a variation of teaching methodology based on the content to be taught and the audience to be addressed. Application of different instructional tools, devices, techniques, materials, etc. is essential for effective teaching because with the comprehension of efficacious teaching objectives the teachers' task and teachers' satisfaction can be achieved (Harper and Chen, 2004). Every student or trainee wants to become more competent, to gather more knowledge about the subject of concern, and integrate his learning into his professional life so that he/she can deliver the best in his/her respective field by way of contributing innovative techniques and ideas. If the student had been taught the concepts in a manner that stays lifelong with the student, then with their cognitive ability and integrated knowledge and skills, one can develop his critical thinking skills and answer the problems faced in a dynamic and challenging profession (Su Gao, 2021). Pedagogical approaches like behaviorism, constructivism, social constructivism, and liberationist are used by the teachers for transferring knowledge to different audiences. For the teacher in every class, there is a mix of students with different learning profiles. For a teacher, before planning the session delivery method it is essential to understand these learning profiles. A teacher may encounter invisible students, some will be underachievers, some may have a unique potential, some may be unmotivated students, some may be prime learners, and some may learn only when they feel it is required (Monoranjan Bhowmik, 2013). With these different sets of students, a teacher needs to follow the pedagogy through which the need of each student can be addressed and information could be transmitted effectively to every student. The teaching pedagogy is applied in a manner to engage all learner profiles and build motivation and confidence in the classroom.

9.7 CHANGES IN TEACHING PEDAGOGY

There are many changes in the ways education is delivered from the days of the gurukul system to modern education. Not only has the education imparted changed but the students have also changed. Previously, the education of girls and women was not appreciated but since the last 100-plus years, educational institutions are now open to women and the importance of their education is generally understood. Statistics of last 50 years bear witness to the increasing composition of female students admitted to schools. A proper curriculum is designed and followed in the schools. Students are grouped according to their age and capability. Professional teachers are hired, and an atmosphere of professionalism is being developed. Moreover, the option of day boarding and boarding schools is also offered to the parents. Students are taught various disciplines and can choose among the various disciplines to become specialized in one particular area (White, 2001). The world has witnessed the change in providing course material from the introduction of books to providing notes in digital format. With the advent of technology, increased demand in the market and changing attitudes, knowledge to be delivered, and the skills needed have also changed (Swartz, 2003). There are a few new trends in education that teachers should think about what and how to teach in the class.

1) Amplified diversity: today a teacher faces the diversified students in the form of child and adult education. Skill based teaching, critical tecing skills, and other diversified techniques need to be imparted by the educator to meet the challenge of diversified trainees. Although it has boosted the teaching career and provided more opportunities in the market.

2) Adaption of new instructional technology: from the chalkboard to smart classes, the instructional technology is changing day by day. A teacher has to get acquainted with these technologies so that true teaching with improved technological assistance can be provided to the students.

3) Increased accountability: today not only students are assessed as to how well they have learned but teachers are also assessed (sometimes by the students) to understand how well they are teaching their students. Teachers are answerable to the parents for the learning of the students, hence, the education system has become more accountable.

4) Raised in professionalism: increased accountability has also increased professionalism. Now the focus is more on quality, establishing higher standards, and continuous improvement.

In the 21st century with advanced technology and digitalization becoming a normal and important part of our life, education is majorly affected by it. The way digitalization is integrated into the teaching pedagogy has changed the ways learning is taking place. The ten major changes are:

1. Inclusion of e-learning in academic programs.
2. The concept of 'teacherpreneurship' has encouraged teachers to create and share their ideas over the internet.
3. Accessibility of all the material on the internet. Here the role of the teacher has changed by making the student learn and understand how to access reliable information and how to segregate the fake and real data.
4. Availability of the application over the phone for easy 24/7 access.
5. Increased mobility due to e-books.
6. Access of education to everyone irrespective of connected and remote areas.
7. Use of digital gaming for quick and effective learning.

9.8 EDUCATION AFTER COVID-19

With the outburst of the COVID-19 pandemic, the whole world was forced to implement lockdown, which meant no schools, no traveling, no offices, no social gathering, etc. After around 100 years, the world population faced the situation where front-line workers had to serve nations at the risk of their life, fighting against this pandemic and saving the lives of countless people. Being the 21st century and due to technological advancement, the good part was that people were connected through digital modes. In fact, for many companies, it was possible to entirely resume work through digital modes, and a new culture of work from home was born. Except for the professions which required work on machines and equipment not available at home, most other organizations started work through digital connectivity.

In these circumstances, offering knowledge and making the youth of the nation's future ready was the biggest challenge for the education sector at all levels of education. Digitalization made it possible for teachers to deliver education and transmit knowledge to the students without incurring any loss of learning (Neetin Prabhu, 2021). Although learning from the distance mode, online video lectures, and online availability, of course, material existed, but it was limited for distance education. After COVID-19 the regular mode of delivering education and teaching pedagogy has changed, as everything is now concentrated on the digital mode of communication (Vijayan, 2021). When changes are made gradually, it's much easier to adapt and the results are usually better. Overnight changes, however, are disruptive and result in many challenges. The same happened with online, or virtual, teaching (Nurul Ashikin Izhar, 2021). Educators and learners were not ready for this sudden change, hence there were many complications; hindrance was faced by both of them (Hashim, 2016). Moreover, in a country like India, network connectivity and electronic devices, etc. are not effortlessly accessible to everyone in every region.

9.8.1 Challenges Faced by Educators

1) Absence of digital skills – shifting to the digital mode of communication for transferring knowledge might be easy for young teachers, but it was a major challenge for the

senior teachers. The use of the internet and different platforms to conduct the classes was already a difficult task for the teacher; along with that they then needed to be trained in dealing with students virtually, using tools like a whiteboard, screen-sharing, screen control, recording, etc. It became essential for teachers to be well versed in using IT tools (overnight!) (Zalat and Hameed, 2021). Using IT effectively in teaching was a great challenge for the teachers.

2) Difficulty in supervision – during online classes, it is difficult for educators to supervise every student whether they are attentive in the class or not, compared to traditional classes due to limited screen size.

3) Difficulty in controlling of class – misuse of IT skills by the students disturbs the class focus and flow. Moreover, related to IT skills, because students are more comfortable with technology than teachers, the teachers would spend valuable teaching time and effort on their own IT skills rather than on their core subjects.

4) Difficulty in evaluation – in the offline classroom, it is easy for teachers to evaluate and monitor while conducting the exam or real-time assignment but with virtual online exams, monitoring is difficult, hence the real picture of students' learning is not clear.

5) Smooth Internet connectivity – although digitalization is the need of the hour and there are many companies offering various internet services, still smooth internet connectivity is not available all the time. Due to poor connectivity, it is always not possible to have smooth, undisturbed online sessions. Many times, students and teachers may not be switching on the camera due to poor network and classes without visualization do not create the necessary impact in the lecture to boost learning.

9.8.2 Challenges Faced by Students

1) Low or poor internet connectivity – as already stated, although digitalization is promoted by the government in India, we face a lack of networking infrastructure. So, poor internet connectivity is a major hindrance for the student to attend online classes.

2) Difficulty following the instruction of teacher due to lack of personal connection – during offline classes, students personally connect with the teachers and want to build image and confidence in front of the teacher. Due to limited screen size and poor connectivity, students feel disconnected from the teacher, and many times it is difficult to follow the instructions properly.

3) Lack of IT skills – although most (young) students are quick learners when it comes to technology, a certain percentage of students have difficulty learning/using technology. Sometimes they also resist learning because post-learning they have to perform and the expectations keep increasing day by day. Also, the student feels overwhelmed thinking that due to IT connectivity they must be available 24/7 for schools or educational institutions.

4) Lack of concentration – watching the screen and maintaining concentration toward it for a continuous 40–60 minutes is not easy for students. They tend to lose their concentration every 15 minutes. Moreover, increased screen time affects their eyes and they need breaks to relax their eye muscles.

5) Feelings of isolation and boredom – when students come to offline class, they enter into the different learning environments, they meet their friends with whom they can share their thoughts, feelings, ideas, etc. In the online mode, their environment remains the same, there is no newness in the environment, and they meet their friends and colleagues officially in the online sessions where they have to be formal, and an absence of connectivity can be felt.

9.9 DIGITALIZATION IS A BLESSING DURING COVID-19 WITH MANY CHALLENGES

Without digitalization, we cannot imagine how the world could have existed during the pandemic. People now understand that it is not compulsory to have a separate infrastructure to work for every kind of profession (Netta Iivari, 2020). Many jobs can work on virtual platforms, having meetings virtually, and store the data and information digitally. Following the same pattern for the long term will help to solve many environmental issues. For instance, since people don't need to travel to offices as much, there will be less pollution, reduced deforestation for paper and the construction of buildings leading to wildlife protection and creating an ecological balance, etc.

Transfer of knowledge through virtual mode benefits students to stay at their homes, avoiding unnecessary traffic issues, wastage of time due to travel, and access to the remote and rural areas. Digitalization had made it possible to create virtual classrooms, sharing the notes virtually, submitting the assignment within the time frame, access to digital libraries, video lectures, chat facilities, etc. (Tudorel Toader, 2021). Many applications have been developed to facilitate education tools easily on the phones. Students and teachers both can use various platforms to communicate, share notes, deliver lectures, make evaluations, etc. (Marko Teräs, 2020). So digitalization is a blessing in disguise during this pandemic situation where the transfer of knowledge, learning was not stopped. It helps teachers to understand the methods in which the new learners can be taught with the use of the latest technology.

9.10 EVOLUTION OF DIGITAL PEDAGOGY

Digital pedagogy explains integrating digital technology into the learning process. It is about how to use the ICT tools and which ICT tools are to be used by the teachers for teaching. There are many tools which can be applied in explaining topics to the students, but what exact tools should be used in which concept can be judged only by a teacher (Janne Olavi Väätäjä, 2021). Digitalization in pedagogy is the result of upgrading technology in every sector. Simply teaching the IT skills to students for employability is not enough; rather, an integration of these IT skills for an effective learning process is also essential (Sadiku, 2016). Post liberalization and opening up of the Indian economy, major changes took place in work patterns, more job opportunities, and more diversified fields, inclusion of technology, and improved information and communication technology (Rebecca Jesson, 2018). During the last two decades, the ICT sector had rapid development where network connectivity is available to every person based on the economic cost.

Internet penetration in India from 2007–2021 (Table 9.1)

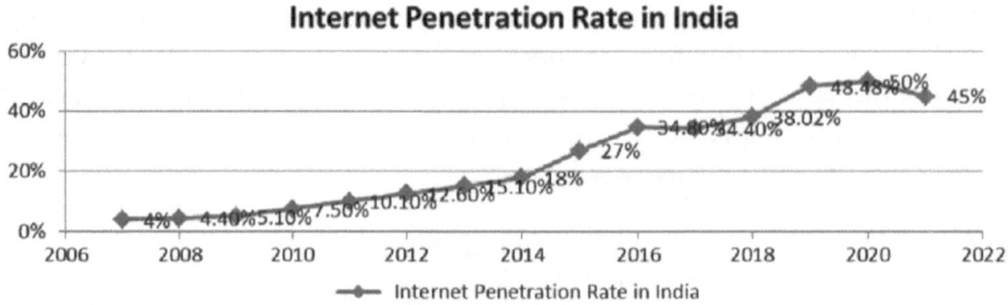

From the above figure, we can easily understand the way the internet had penetrated the Indian economy. Today the number of mobile users using internet facilities had tremendously increased. This has benefited the education sector by providing easily accessed information to all regions.

Indian IT Output Levels and Growth Rate (1991–2015)

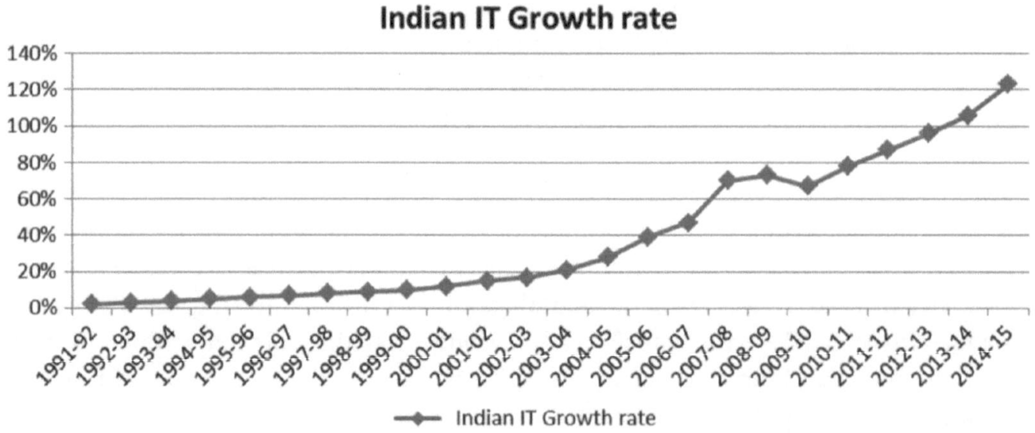

Before understanding the growth of ICT in the education sector let's first understand the growth of Information and Communication Technology in India. In the graph below it can be observed that before 1991 the usage of IT tools was very negligible but post 1991–92, there was the introduction of IT tools in but acceptability was very less. The post year 2000–01 the penetration and acceptability of IT had increased and the graph is rising in the increasing rate. This depicts the population as becoming tech say and getting comfortable in using technology not only professionally but also in their personal life. So there is no way that the education sector has escaped from IT skills it is the sector that divulges knowledge and makes people ready to work with technology. The sector itself has also combined traditional learning with innovative technology-based learning to facilitate teachers and students for better erudition.

TABLE 9.1 Internet Penetration Rate in India from 2007–2021

Year	Internet Penetration Rate in India
2007	4%
2008	4.40%
2009	5.10%
2010	7.50%
2011	10.10%
2012	12.60%
2013	15.10%
2014	18%
2015	27%
2016	34.80%
2017	34.40%
2018	38.02%
2019	48.48%
2020	50%
2021	45%

Contributions of the IT Sector from 1950 to 2010

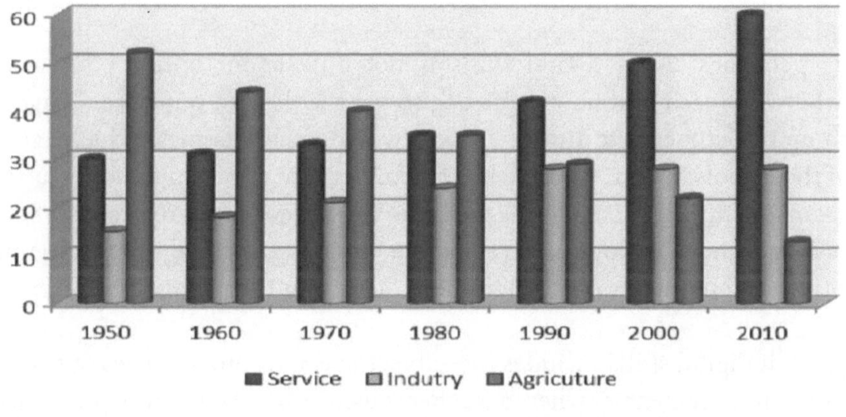

The above graph gives a very clear picture of the rise of contributions of ICT in the service industry. Initial stage of technological development was related to industrial development only but in the later phase with computerization, every segment in the country started espousing it. When talking about the education sector, it is a noble service provided to the nation and plays a major role in nation-building. Teachers and educators were expected to use different tools for their classroom instruction, exchange of notes, and conduct the examination. The use of audiovisual aids like OHP, projectors, slides, e-mails, etc. have become very common in schools and colleges. This helped teachers to prepare their lectures in an interactive and interesting way, and students were able to grasp the knowledge more easily as visual memories are stronger than our hearing memory. By the year 2000, there were no educational institutions in rural or urban areas that were not using any kind of ICT tools and sticking to the traditional way of teaching.

Share of Students with Access to Educational Resources by Household Income

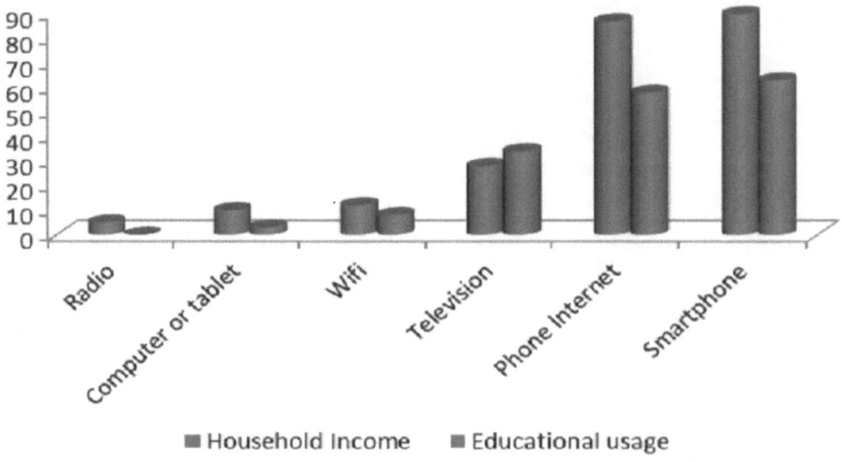

From the above figure, it can be concluded that there is more usage of smartphones and the internet through phones. These statistics have been leveraged by the IT professionals who developed a user-friendly application for easy learning, communication, and exchange of notes. So digital skills today are not limited only to the effective usage of computers, instead, it has expanded to the use of smartphones for effective learning and monitoring.

Teaching pedagogy using digital skills is not only about online education, it also describes what you teach and how you teach to get the desired outcome. Usage of digital tools opens up our students to the new world which itself has many challenges. Careful selection of those tools in connection with the curriculum to be taught for creating a long-term impact on the minds of students is the creativity a teacher has to exercise. So, teaching may seem easier with technological development, but it is a challenging task to understand all the positives and negatives of ICT tools and designing the pedagogical framework (see Table 9.2).

Teaching with digital skills includes the subject-specific competencies of teachers combined with digital competencies where a teacher is using digital resources embedded in their teaching and learning. Here teachers use these digital resources for evaluation and assessment and also empower learners to gain transversal and subject-specific competencies.

9.11 TYPES OF ICT TOOLS AVAILABLE FOR TEACHING

There are many information and communication tools available for teachers for blended learning. These tools can be categorized as follows:

1) Informative tools – there is a large amount of information available in the form of text, images, sounds, or videos on various digital platforms. Nowadays accessing e-libraries, online journals, online newspapers, and other information is trending. It is more convenient to search and read the information at your place according to your schedule. These information portals are open 24/7 hence the user is not time-bound

TABLE 9.2 Areas and Competencies of Educators in the Digital Inclusive Pedagogy

Professional engagement	Educators' professional competencies
Organizational communication	
Professional collaboration	
Reflective practice	
Digital CPD	
Digital competencies	
Subject-specific competencies	
Digital resources	Educators' pedagogic competencies
Selecting	
Creating and modifying	
Managing, protecting, and sharing	
Teaching and learning	
Teaching	
Guidance	
Collaborative learning	
Self-regulated learning	
Empowering learners	
Differentiation and personalization	
Accessibility and inclusion	
Actively engaging learners	
Assessment	
Assessment strategies	
Analyzing evidence	
Feedback and planning	
Transversal competencies	Learners' competencies
Facilitating learners' digital competencies	
Information and media literacy	
Communication	
Content creation	
Responsible use	
Problem-solving	
Subject-specific competencies	

as far as accessing information, and it is also updated instantly by the websites/portal holders.

2) Situating tools: there are many web-based applications that allows teachers to create simulation games for students, putting them under the real situation, so they can react and learn from it. Customization of such applications could also be done by professionals if an institution wants to have it specifically for their students. This type of learning helps the student apply their conceptual skills to real practice, making learning more vigorous, inventive, collaborative, and nourishing.

3) Constructive tools: these are simple applications available online and offline to develop teachers' lectures. This can be the usage of MS office for Word, PowerPoint, and Excel, creating diagrams, videos, and audios. Once prepared the same can be utilized again after updating, if required, and also can be shared easily with students.

4) Communicative tools: for sharing any notes or informational e-mails, institution websites, whatsapp, messenger, telegraph, SMS, etc. tools are used by the teachers. These are easy and reliable methods for communicating any sort of information to students. The online mode of teaching platforms like Microsoft Office, Zoom, Google Classroom, etc. had served as an important tool for conducting classes through video conferencing, assessment, and sharing assignments.

5) Collaborative tools: if any project or assignment has to be done in collaboration, and working on the same document is required, usage of interactive whiteboard, facility of screen-sharing, various discussion forums; usage and creation of blogs; use of LinkedIn for professional networking; sharing and saving documents/notes on the Cloud or Google Drive, etc. can be done easily due to these user-friendly two-way tools available on the internet.

9.12 DIGITAL LITERACY SKILLS REQUIRED FOR TEACHERS

A teacher will be declared digitally literate when he/she is able to traverse, gauge, and construct using all forms of digital technologies. Prior to becoming digitally literate, it is essential that teachers should be comfortable with applied technical skills, collaboration and integration skills, social and cultural skills, and critical thinking (Yazon, 2019) (see Figure 9.5).

FIGURE 9.5 Digital literacy competancies for teachers. Gurukul training methodology.

1) Critical thinking skills: critical thinking skills refers to when a student can collect relevant information, analyze it, observes the pattern, can connect all possible dots, evaluate the data, and is capable of identifying and developing meaningful information from the data collected. When there is lots of information already available on the internet, and knowledge can be gained from it, the question that now arises is 'what is the role of a teacher?' It is a challenging task for a teacher to teach when students can counter the questions using information they accessed from the internet. So, it is essential that a teacher should plan their lecture in a way to encourage students to search for all sorts of relevant information available from the World Wide Web – the sort of information that creates connections – then analyze it, and present the data in an informative manner. This will enable the student to segregate between real and fake data, how to access reliable sources, question themselves like who is the source, for whom data is mentioned, who is affected, what would I want to know, what is important, searching for the evidence, looking for validation, asking for when and why for the data searched, etc. Hence it is essential for the teacher to design projects using his critical thinking, so that it should develop the critical thinking skills of the students.
2) Online safety skills: the Internet is a tool full of benefits and lots of risks: risk of theft, risk of losing data, risk of privacy invasion, risk of leaking confidential information; so, the teacher should be well versed in the potential risks of using digital skills or teaching through online mode. Moreover, students should be taught about the measures to minimize risks and how to handle them. Virus protection, cyberbullying, plagiarism, and copyright infringement are a few topics of which teachers and students both should be aware.
3) Digital culture: apart from social culture, the next thing which needs to be framed ethically and learned by the older generation is digital culture. Today, the internet and smartphones with lots of social apps has enabled people to connect with others all over the globe: share their views, opinions, achievements, experiences, pictures, stories; form groups of similar interest, start a revolution on certain issues globally, etc. It is also somewhat creating the privacy issue. While teaching, an educator needs to understand the digital culture of which this young generation are major participants. Apart from opposing, it requires them to learn and divert their energies toward positive and constructive formats of learning in the digital environment. When technology is used in the right path, it is a blessing; otherwise, if left unattended, it may create disaster.
4) Collaboration and creativity: there are many web-based tools and smartphone-based applications available for teachers to blend their subject knowledge digitally. For example, preparing lecture videos and sharing them publically on Youtube channels benefit the very large audience of those who want to gain knowledge; creating blogs where students can be motivated for writing on various topics which can be shared publically to receive comments, etc. are few ways the teacher can use their creativity and integrate traditional learning with digital skills.

5) Finding information: the internet is an ocean of information, and with one click, there are lots of websites providing the information on the topic searched. With growing digitalization, one word that is common nowadays is 'trending'. When a video is 'trending' it means it is being followed by many viewers, but quite often the trending is manufactured using search engine optimization (SEO) tools and techniques. In this case, anything that goes viral might be fake. Before starting any search, the following points should be taken into consideration:

    a) Think properly and be clear about your search.
    b) Apart from a random search on search engines like Google, Bing, etc., there are many reliable online libraries providing lots of information.
    c) Engage in a deep search and don't get stuck on the first few results.
    d) Try various permutations and combinations of words while searching.
    e) Choose your keywords carefully.
    f) Evaluate the results by checking the website, address, check the reliability of the source, study beyond the taglines, understand if there is any alteration in images or information, and cross-check with other sources.
    g) Check the facts on your unbiased judgment.

6) Communication skills and netiquette: the internet is all about communication, but most people are bad at communicating on the internet. Effective communication skills will help to create a strong digital imprint on the internet and ensure online safety. Platforms like soil networking sites. Real time messaging service, video/audio conferencing, live streaming, blogs, vlogs, e-mail, chat rooms, forums, etc. are designed to fulfill the separate social and professional needs. Each platform has certain protocols to be followed, so ethics and etiquette while being online are a must for teachers and students.

7) Functional skills: access to the technology and using the technology depend on how comfortably it can be learned and used by teachers and students. It requires the availability of resources and knowledge on both ends.

9.13 BENEFITS OF DIGITAL SKILLS IN TEACHING PEDAGOGY FOR TEACHERS

1) Easy preparation of lecture notes: with the usage of tools available on the internet and built-in in the software, it becomes easy for the teacher to prepare the lectures. Effective use of these tools helps teachers in quick preparation, display of diagrams, etc.

2) Interactive and enticing lectures can be prepared: usage of animation, the inclusion of audio/video notes, generating hyperlinks for easy navigation, clipart, color combinations, etc. make the lectures attractive and interesting for the students and help

teachers by keeping the students' focus on the lecture. It also reduces the boredom of just listening to the teachers.

3) Creation and maintenance of databases: once a lecture is prepared, it can be saved for the future and reused with a different audience.

4) Easy to carry: as all the notes, presentations, audio, and video can be saved in a small device called a 'pen drive', it is easy to carry all the lectures in the pocket. With the increased usage of the internet and cloud management, carrying the pen drive is also not required. With amplified connectivity resources the lecture can be accessed anywhere at any time globally.

5) Sharing of information: receiving the assignments digitally through e-mails or institutional websites, or the real-time sharing through the Cloud or drives have made sharing of information possible and easy.

6) Increased audience: having a good quality camera on every smartphone has facilitated teachers to prepare video lectures with blended PowerPoint presentations or other tools which can be uploaded through Youtube channels. If these lectures are made public, they can be accessed by anyone worldwide. In this way, the learning is not limited to a few students; it can be accessed by a large number of people.

7) Conducting online quizzes and assessments: customized quizzes and different forms of assessment can be prepared through which regular evaluation of students is possible. Marking or assigning marks and providing feedback is also possible without much effort as these platforms facilitate automatic features for it.

8) Easy maintenance of database: all student-related information like class assessment, attendance, exams marks, etc. can be maintained effortlessly.

9) Increased Opportunities: the digital education market is growing very rapidly and there are many players coming up with different applications or platforms where online education on specific subjects is possible. The job opportunities for the content creator, evaluator, and online teaching are increasing.

9.14 BENEFITS FOR DIGITAL SKILLS IN TEACHING PEDAGOGY FOR STUDENTS

1) Reduced mobility and time and energy saving: access to online libraries, video lectures, real-time classes, etc. enable the student to reduce their traveling time and save energy for learning.

2) Learning of digital platforms: inclusion of digital skills in teaching makes students acquainted with the various digital platforms and how to use them. Many of the applications like Word, PowerPoint, Excel and skills like effective search through proper keywords on search engines, the Clouds, and drives are used by organizations

in their regular working. If students are already aware of these technologies, then separate classes are not required for the students.

3) Reduced boredom: the traditional way of teaching were monotonous and required special effort from students to maintain focus in the session. But the blended education and technology method offers an exciting and easy way to learn. Studies become interesting for students and concepts are easily grasped and retained by the students.
4) Cultivate accountability among students: real-time monitoring and evaluation through auto-generated reports which are transparent to the students enable them to measure their performance and work upon the solutions to improve it.
5) Increase engagement: using digital tools, platforms, technology by students engage them in learning, as they have ample information available on the internet to study, various technical user-friendly tools to aid in learning, discussion forums to share their views, social platforms to create connections, etc.

9.15 CONCLUSION

Teaching plays an important role in our lives, from ancient times to the present era we have been blessed with teachers who shared their wisdom, transferred knowledge, and built capable citizens for the building of the nation. The ever-growing change which we observed today is made by the people who had been taught by some of the great teachers. Any kind of technical growth is the outcome of hard work, innovation, and creativity of the tech-savvy and developers. But to become competent in using the technical skills and able to invent new technology they had undergone formal education where effective teaching pedagogy of teachers had erected their learning and abetted in knowledge creation. So the role of a teacher with digital skills or not can be ignored. With changing time and dynamic learning environments adoption of digital skills by teachers and integrating it with traditional learning is the major challenge for them. But as it said a teacher does not hesitate in using any method for making their student educated. So today teachers in every field have conquered the use of budding technology in their teaching methods. Increased awareness and knowledge of digital skills has raised the demand for specific tools or applications or software to meet the everyday demands of teaching and making the teaching more collaborative, stimulating, transparent, and effective.

REFERENCES

(n.d.). Retrieved September 4, 2021, from https://courses.lumenlearning.com/: https://courses.lumenlearning.com/boundless-psychology/chapter/introduction-to-learning/

Baker-Rush, M. L., Pabst, A. (2021). Fear in Interprofessional Simulation: The role of psychology and behaviorism in student participation and learning. *Journal of Interprofessional Education & Practice*, *24*, 1–18.

Boyan Bontchev, D. V.-P. (2018). Playing styles based on experiential learning theory. *Computers in Human Behavior*, *85*, 319–328.

Chaliha, A., Borah, S., (2016). *Foundation Of Education*. Jorhat: Bidya Bhavan.

Chutia, M. (2020, September). Growth and development of education in india during the British period in a historical perspective. *International Journal of Management (IJM)*, *11*(9), 1464–1470.

Cuenca, A. (2010). Self-study research: Surfacing the art of pedagogy in teacher education. *Journal of Inquiry & Action in Education*, *3*(2), 15–29.

Enliang Yan, C. Y. (2021). Incremental concept cognitive learning based on three-way partial order structure. *Knowledge-Based Systems*, *220*, 1–11.

Frederick, A. G. (2016). A comparative study between gurukul system and western system of education. *IRA Acdemico Reserach*, *3*(1), https://research-advances.org/index.php/IJEMS/article/view/116 (accessed on 8th July).

Guillaume Fontaine, S. C. (2021). A theory-based adaptive E-learning program aimed at increasing intentions to provide brief behavior change counseling: Randomized controlled trial. *Nurse Education Today*, *107*, 105–112.

Harper, K. C., Chen, K. (2004). Distance learning, virtual classrooms, and teaching pedagogy in the Internet environment. *Technology in Society*, *26*(4), 585–598.

Hashim, H. N. (2016). Teachers' challenges in the implementation of frog virtual learning environment in the classrooms. *Asia Pacific Journal of Educators and Education*, *31*, 115–129.

Jan Moons, C. D. (2013). The design and pilot evaluation of an interactive learning environment for introductory programming influenced by cognitive load theory and constructivism. *Computers & Education*, *60*(1), 368–384.

John Dunlosky, K. A. (2013). Improving students' learning with effective learning techniques: Promising directions from cognitive and educational psychology. *APS Associaton for Psychological Science*, *14*, 4–58.

Karve, S. K. (2013). Comparative study of ancient gurukul system and the new trends of guru-shishya parampara. *American International Journal of Research in Humanities, Arts and Social Sciences*, *13*(140), 81–84.

Kayalar, F. (2017). The effects of auditory learning strategy on learning skills of language learners. *IOSR Journal of Humanities And Social Science (IOSR-JHSS)*, *22*, 4–10.

Kibble, D. K. (2016, February 4). Learning theories 101: Application to everyday teaching and scholarship. *Advances in Physiology Education*, *40* (1), 17–25.

Luniya, B. N. (2012). *Evolution of Indian Culture: From Earliest Times to the Present Day*. Lakshmi Narain Agarwal.

Madsen, S.R., Wilson, I. K. (2012). Humanistic theory of learning: Maslow. In: Seel N.M. (eds). *Encyclopedia of the Sciences of Learning*. Springer.

Marko Teräs, J. S. (2020, October). Post-Covid-19 education and education technology 'solutionism': A seller's market. *Postdigital Science and Education*, *2*, 863–878.

Mehmet Kesim, H. A. (2015). A theoretical analysis of moocs types from a perspective of learning theories. *Procedia: Social and Behavioral Sciences*, *186*, 15–19.

Mitchell, R. (2018). Experts warn play time is 'disappearing' as emphasis is placed on performance and tests. *The West Australian*, *4*, 1014.

Monoranjan Bhowmik, B. B. (2013). Role of pedagogy in effective teaching. *Basic Research Journal of Education Research and Review*, *2*(1), 01–05.

Mookerjee, S. P. (n.d.). Education in British India. *The Annals of the American Academy of Political and Social Science*, *233*, 30–38.

Mukherji, S. N. (1964). *Education in India-Today and Tomorrow*. Acharya Book Depot Baroda.

Navaneedhan, C. G. (2011). Reflective teaching pedagogy as an innovative approach in teacher education through open and distance learning. *Academic Journals*, *3*(12), 331–335.

Neetin Prabhu, C. L. (2021). Building better education after COVID-19: The 2021 Canadian ophthalmology residency match. *Canadian Journal of Ophthalmology*, *2*, 52–54.

Netta Iivari, S. S.-O. (2020). Digital transformation of everyday life: How COVID-19 pandemic transformed the basic education of the young generation and why information management research should care. *International Journal of Information Management*, *55*, 1–6.

Nurul Ashikin Izhar, Y. M.-d. (2021, February 24). Teaching in the time of Covid-19: The challenges faced by teachers in initiating online class sessions. *International Journal of Academic Research in Business and Social Science*, *11*(2), 1294–1306.

Proctor, K. R., Niemeyer, R. E. (2020). Retrofitting social learning theory with contemporary understandings of learning and memory derived from cognitive psychology and neuroscience. *Journal of Criminal Justice*, *66*, 1–16.

Raiyn, J. (2016, 115–121). The role of visual learning in improving students' high-order thinking skills. *Journal of Education and Practice*, *7*(24), 115–121.

Rebecca Jesson, S. M. (2018). Improving achievement using digital pedagogy: Impact of a research-practice partnership in New Zealand. *Journal of Research on Technology in Education*, *50*(3), 183–199.

Robinson, S. K. (2012). *Out of Our Minds: Learning to be Creative*. Wiley Online Library.

Sadiku, M. (2016). Digital pedagogy. *International Journal of Trend in Scientific Research and Development*, *3*, 801–802.

Sanmugarevathi, M. (2020). *Application of Humanism Theory in the Teaching Approach.*

Saranya Kumar, S. (2016). Gurukula system of education in ancient times. *Business Sciences International Research Journal*, *4*(2), 44–45.

Saravanakumar, D. (2019). Unit 1 Organisational Behaviour. *Alagappa university SIM Mode Book*, *53*. Organisation Behavior.

Selvamani, P. (2019). Gurukul system: An ancient educational system of India. *International Journal of Applied Social Science*, *6*(6), 1620–1622.

Shrivastava, A. (2018). Using connectivism theory and technology for knowledge creation in cross-cultural communication. *Research in Learning Technology*, *26*, 1–16.

Sitthiphong Suwannaphisit, C. A. (2021). Expectations on online orthopedic course using constructivism theory: A cross-sectional study among medical students. *Annals of Medicine and Surgery*, *67*.

Soni, B. (2018). *A Case Study on Gurukul System of Education*. SSRN.

Soni, B. K., Trivedi, J. C. (2018). A case study on gurukul system of education a contemporary approach by gotirth vidyapeeth. *Rajagiri Management Journal*, *12*(2), 245–247.

Su Gao, N. D. (2021). Mapping and reflecting on the integration of the components of pedagogical content knowledge (PCK) for teaching natural selection: A case study of an experienced middle-school science teacher. *Teaching and Teacher Education*, *107*.

Swartz, E. (2003). Teaching white preservice teachers: Pedagogy for change. *Urban Education*, *38*(3).

Tomar, D. (2014). A study of effective learning strategies in relation to intelligence level across the science and arts academic streams of secondary level. *IOSR Journal of Research & Method in Education*, 41–50.

Tudorel Toader, M. S. (2021, June 5). Effects of digitalisation on higher education in a sustainable development framework: Online learning challenges during the COVID-19 pandemic. *MDPI, 13*.

Väätäjä, J. O., Ruokamo, H. (2021, March). Conceptualizing dimensions and a model for digital pedagogy. *Journal of Pacific Rim Psychology*, *15*, 1–12.

Vijayan, R. (2021, July 12). Teaching and learning during the COVID-19 pandemic: A topic modeling study. *MPDI*, 11, 1–15.

White, B. (2001). *Ancient Models, Modern Teachers, and Moral Education: Choosing a Peaceful Pedagogy*. Retrieved September 22, 2021, from https://www.tcrecord.org: https://www.tcrecord.org/Content.asp?ContentId=10734

Yazon, A. (2019). Digital literacy, digital competence and research productivity of educators. *Digital Literacy, Digital Competence and Research Productivity of Educators*, *7*, 1734–1743.

Yun-Jeong Oh, H.-Y. K. (2021). Effects of a transformative learning theory-based debriefing in simulation: A randomized trial,. *Nurse Education in Practice*, *50*, 1–6.

Zalat, M. M., Hameed, M. S. (2021, March 26). *The Experiences, Challenges, and Acceptance of e-learning as a Tool for Teaching during the COVID-19 Pandemic among University Medical Staff*. Retrieved September 22, 2021, from www.plos.org: https://journals.plos.org/plosone/article?id=10.1371/journal.pone.0248758

CHAPTER 10

National Education Policy-2020

Prospects for Teacher Education Institutions

Anuradha Agnihotri

CONTENTS

DOI: 10.1201/9781003254942-10

10.1 INTRODUCTION

The system of education in India has been a subject of discussion ever since the Wood's dispatch, the Magna Carta of Education implemented by the British government in 1854 (pre-independence times), resulting in westernization of education and opening of many universities in India such as at Calcutta, Bombay, and Madras. University Education Commission (1948–49) was the first education commission of its kind after independent India. This was followed by several Education Commissions, prime being the Secondary Education Commission (1952–53), also known as the Mudaliar Commission; the Indian Education Commission (1964–66), also known as the Kothari Education Commission; National Education Policy (1986, revised in 1992), and the National Knowledge Commission (2005, defunct since 2014).

The present education policy, National Education Policy, that first came in 2019 that was immediately withdrawn to incorporate some drastic changes to it, came again in 2020. A comparison of NEP-2020 with the previous education policies of 1986 and 1992 and also with the reports of the iconic Indian Education Commission (1964–66) reveals many similarities in the issues, prospects, and even the suggested reforms, except for the focus on 21st century skills, such as IT skills, soft skills, and hands-on experience. It does not mean that the present policy is just a mere addition of some focal points to the previous policies, like an old wine in a new bottle. NEP-2020 is a major shift from the old school of thoughts in terms of the structure, implementation, and vision of various programs of the school and higher education system in India. It has brought into the limelight the major concerns of our country, such as the need for multidisciplinary and interdisciplinary institutions suited to the location (urban vs rural), nature of education (school education vs higher education), types of courses (academic vs professional), and focus on quality of institutions vs quantity (industrial training institutes (ITIs) vs Indian institutes of technology (ITTs)).

10.2 STRENGTHENING TEACHER EDUCATION THROUGH NEP-2020

There are grave concerns about the quality of teacher education in India. Of late, these have become the focal points of discussion on all education forums. The value of NEP-2020 will be diluted if no reforms are brought to the present system of teacher education which has deteriorated in the hands of poorly performing teacher education institutions (TEIs). It would be worthwhile to see what changes ensue in the quality and status of teacher education programs, bringing quality through continuous professional development programs, upliftment in the status of teachers thereby enhancing the prestige of teachers in our society. NEP-2020 has advocated that these issues have to be dealt with on a priority basis before the implementation of policy decisions. There is a dire need to bring drastic structural reforms and not just superficial changes.

If we carefully read NEP-2020 paragraphs 15.1–15.11, we would find strong criticism of the follies of the current teacher education system. The policy envisages that National Council for Teacher Education (NCTE) needs to be replaced before making any attempts to reform TEIs in our country. This has been mainly due to the irregularities in the functioning of NCTE for a long time, which has resulted in the mushrooming of TEIs across nation and the production of poor quality teachers. Many of these TEIs are not concerned with the quality of education and are just opened for business' sake. The responsibilities of NCTE need to be transferred to other regulatory bodies of higher education, such as UGC, NCERT, or CBSE. Or perhaps a new, autonomous agency needs to be set up by the government. Such a daring effort on the part of the government would speak volumes about its will and determination to bring improvements in teacher education.

During the past two decades, the quality of teacher education has deteriorated as has been expressed by different national committees that have mentioned this fact quite emphatically. NEP-2020 has come as a fresh wave that has added new dimensions to the present system of education, suggesting possible reforms from the pre-primary to higher education sectors. We need to reflect upon the causes of the present shattering status of teacher education before moving ahead, lest in a hurry we either repeat the same mistakes or discard the assets of the existing system. We need to learn lessons from mistakes of the past for a better future. The blame game among stakeholders is pointless, as each one has more or less contributed toward the present status of teacher education. If large numbers of TEIs have not performed up to the mark then the governing and apex bodies too have slackened in performing the supervision and guidance tasks. This needs an analysis of their past performance.

It is an established fact that the present system of teacher education is deeply problematic with all possible forms of malpractices prevalent in it. It is more of a business, than a concern for the nation; middlemen thrive in it. Presently, TEIs are allowed to have an annual intake of 100 students for a two-year B.Ed. program. There are many universities which are not even following proper norms related to admission in teacher education courses. Mostly the admissions are based on merit, i.e., by taking into account the marks secured in graduation level; few universities are still conducting entrance tests for the same. But the majority of the TEIs are not adhering to basic norms of admissions.

I have observed that there has been reluctance on the part of the students from the time the B.Ed. course from a one-year to a two-year course (2015). Similarly, the M.Ed. course also become a two-year course. This reluctance is likely due to students (often from less affluent backgrounds) not being able to afford the two-year course fees. The prescribed norms for the fee structure of the university are also followed by these money-minting TEIs, as they charge surplus from the students. The result is that many of the seats lie vacant in such institutions, and they resort to unethical means to bring students to their campuses. The faculty of these institutions are compelled to recruit students from far off, remote and rural areas, under threat of losing their jobs. Many of these institutions have been forced to close due to a lack of students. Moreover, the students who do take admissions are mostly non-attending because they are working side-jobs, which is a gross defiance of the rules.

The staff in such TEIs are hired on a temporary basis, paid meager salaries, and are mostly underqualified. Even the principals are hired on a poor salary, employed merely for the sake of inspections and signatures. Thus, it is not surprising that such poorly functioning TEIs presently need a total transformation, or else they need to be closed down. There is a good possibility that as per the recommendations of NEP-2020 of having 4-year integrated B.A./B.Ed. program, many of the poorly performing TEIs would be forced to shut down due to non-compliance of proper rules and regulations.

We also need to accept yet another overwhelming fact: quality in higher education has deteriorated in every undergraduate (UG) course, be it B.A., B.Sc., or B.Com. This has resulted in lack of content mastery in specialized fields among students before entering the teacher education programs. This gap cannot be filled in TEIs, where 'how to teach' is more in focus compared to 'what to teach'. So, in this transition one thing is sure: if we blindly go in for the generalization of teacher education in composite colleges, then we may end up repeating the same mistakes over and over again. Before we embark upon this new venture, the fact has to be acknowledged that teacher educator preparation as well as teacher preparation is not at all an ordinary stream of studies, and it must be given the status of a professional course; unfortunately, a panorama which has been missing so far in the entire history of teacher education in India. There may be many reasons for the present undesirable scenario, but now we have to move forward with a planned strategy to implement NPE-2020 and bring qualitative improvements in teacher education by making a smooth transition from unproductive teacher education degrees to a meaningful and outcome-based teacher preparation, a vital propelling force needed for the overhaul of the entire education system.

The NEP-2020 focuses on a major aspect that has for so long required our attention toward the education system in India – that is, the quality of teachers that our system produces. The aim is to tread toward a flexible, interdisciplinary progressive society from a rigid discipline-centered education, despite the huge demographics that challenge us. The changes that the policy wishes to bring in our system of education should be seen in the light of creating more thoughtful, engaging, competent, and skill-oriented teachers from the knowledge-oriented teachers. This has been previously reflected in reports of various education commissions.

The Kothari Commission (1964–66) also remarked that it is the quality, competency, and conduct of the teachers that contributes to the development of the nation. NEP (2020) also emphasizes the importance of good quality teachers by saying that it is the teacher who truly shapes the future of our children and nation thereby implying that teachers play the most important role in nation-building by creating high-quality human resources in their classrooms.

Although the vision is strong, there are many hurdles in the way of producing competent teachers due to practical difficulties that the teacher education institutions (TEIs) are facing. As per the Justice J. S. Verma Committee Report (2012), the system of teacher education is in a fragmented state, putting millions of children at risk. The infrastructure is minimal, and the majority of teachers failed to pass the Central Teacher Eligibility Test (CTET), even after completing pre-service teachers training. The challenges that TEIs are facing include poor student enrollment, mushrooming of stand-alone TEIs in the region, and the poor working conditions of teacher educators (hiring them on a temporary basis, poor salaries, absence of promotion, denial of opportunities for professional development and other benefits, relieving teacher educators in the mid-session, lack of training facilities – the list is long). This has resulted in a lack of knowledge, skill sets, professionalism, and commitment on their part.

NEP-2020 blames the poor employment conditions of teachers for the degraded system of teacher education in India. It now attempts to bring systemic reforms to teacher education to attract students toward the noble profession of teaching, which otherwise has been observed to be a compulsion rather than a choice. Here by compulsion, I mean that a majority of the students pursuing pre-service teacher education are either forced by their parents as a result of the prevalent thought that teaching is a convenient and easy profession meant for ladies to manage their family responsibilities after marriage; or they have no other professional degree after their graduation, so they are compelled to pursue B.Ed. NEP-2020 attempts to attract the talented and young minds toward teacher education not out of compulsion but by choice. It proposes many reforms to strengthen the status of teachers and restore the dignity and respect that it has lost over the course of time, with the prospect of eventually attracting the most talented students to choose teaching as their profession.

10.3 PRE-SERVICE TEACHER EDUCATION AS ENVISAGED IN NEP-2020

As per the recommendations of NEP-2020, a National Curriculum Framework for Teacher Education will be drafted, the committee for which has been formed. It will serve as a document to guide all teacher education institutions regarding the teacher training, both at pre-service and in-service levels, for the teachers working in academic, vocational, and special education streams.

NEP-2020 has recommended a four-year integrated B.A./B.Sc./B.Com./B.Ed. degree after +2 stage of schooling, and all stand-alone TEIs have been asked to convert to multidisciplinary institutions by the year 2030. This will be the minimum qualification in order to become school teachers. National Testing Agency (NTA) will conduct an aptitude test for admission to this course. The teacher education institutions, which are well equipped with infrastructure, have competent and qualified staff and can fulfill all the conditions laid down by the recognized bodies like NCTE, UGC, NAAC. Affiliated state/central universities should be given approval to start the 4-year integrated course. This way, the present premier TEIs will be able to contribute toward producing effective prospective teachers, as envisioned in NEP-2020.

All degree colleges and universities have been directed to set up an education department to run B.Ed. (which, so far, have been undertaking Masters in Education courses (M.Ed.) and doctorate degrees only) by collaborating with other departments such as sciences, arts, commerce, languages, psychology, music, physical education, history, philosophy, mathematics, sociology, etc. In addition, these institutions have been directed to focus on quality researches in the field of education to improve the quality of their B.Ed. programs.

This integrated B.A./B.Sc./B.Com./B.Ed. degree will focus on a wide range of pedagogies along with core foundation knowledge-based subjects along with a rich practicum. The contents will include pedagogies for CWSNs (children with special needs) and other diverse learning needs, online teaching, regular classroom teaching utilizing educational technology, collaborative learning, contemporary techniques of evaluation, etc.

The teacher training must include various components such as orientation in theoretical foundation subjects, face-to-face classroom learning, self-directed learning, peer observations, internships, practicum, etc. irrespective of the duration of the teacher education program in order to achieve the learning outcomes. The entire course should be divided into semesters. Instead of the four-month internship that is currently in use, a full one-year internship in a school under the guidance of mentor teachers will be mandatory to give the pupil teachers a thorough knowledge of the inner working of schools. During the internship, the teacher trainees will observe classes taught by experienced school teachers and also by their fellow peers. They will be required to prepare comprehensive reports of their experiences along with their reflections about the internship. These reports will be evaluated during their practical examination. The mentors supervising them will also give their feedback about the performance of the teacher trainee in different aspects, such as thoroughness of content, efforts to make lessons interesting, teaching–learning materials used during their lessons, pupil engagement, activities or experiments conducted during the lessons, different methodologies used to make lesson more interactive, and so on. The idea is that the teacher trainee must be properly familiar with the environment of the school before joining the teaching profession.

NEP talks about sharing of talents and resources among the school clusters in running TEIs. The idea behind formation of school clusters is that the trainees who pass out from TEIs would ultimately serve in the schools itself. So, the need of the hour is that the school clusters may be allowed to set up TEIs and run the courses as per a broad curriculum framework framed by regulatory bodies with a good sense of autonomy with respect to infrastructure and human and material resources.

NEP-2020 recommends that big schools which employ more than 200–300 teachers may be allowed to run TEIs so that they may be able to prepare the teachers as per their requirement and can ensure mastery in entry-level skills in the areas of information and communication technology (ICT), digital pedagogies, communication, and other soft skills. Even the upcoming universities with established infrastructure and academic credentials, which have been outperforming for the past few years, may be allowed to start teacher training courses without any delay.

Detailed guidelines need to be shared regarding structure and nature of a four-year B.Ed. program. The curriculum for this has to be framed in such a way that the students who have enrolled themselves for the course, must be eligible to pursue postgraduate programs in the optional subjects in the four-year course. For this purpose, a well thought out credit transfer based on a holistic curriculum is required.

NEP-2020 advocates that multidisciplinary institutions should replace the present stand-alone TEIs by the year 2030 and they have to become an essential part of higher education. Collaborative prompt policy decisions by NCTE, MOE, state governments, and affiliated universities are required to be taken to convert stand-alone colleges to composite colleges. The existing stand-alone colleges with necessary infrastructure and staff requirements should be allowed to run degree courses such as B.A., B.Sc., and B.Com. to be able to convert into composite colleges in its true essence. However, we should not be in an unprecedented hurry to lead to a demolition drive for the stand-alone TEIs and introduce a new system of four-year integrated teacher education degrees (B.A./B.Ed., B.Sc./B.Ed., B.Com./B.Ed.) in newly defined composite colleges. The faculty of composite institutions must be given training for pedagogical aspects, and teacher education institutions must be well prepared for the content aspect of a particular integrated course.

The training of the faculty should include proper integration of pedagogical aspects with that of the content. For this, the faculty of both fields need to ponder upon the ways to integrate the pedagogy with the content and fix the ratio of credits. This integration must be given careful thought by all the stakeholders of education. There must be proper strategies on sharing the workload between the teachers of pedagogy and those who will be teaching the content part. In order to successfully run the integrated teacher education programs, functional linkages between different departments must be encouraged.

Existing performing institutions should also be given the opportunity to run other courses of teacher education such as integrated B.Ed./M.Ed. program, B.Ed. (special education), N.T.T., E.T.T., D.El.Ed., etc. The NPE-2020 focuses on preparation of special educators (reference is given in policy document), so existing institutions should be given the opportunity to run these different hues of teacher education and allied courses and suitable guidelines, and conditions should also be framed for the same.

As NPE-2020 encourages the use of existing resources on a sharing basis, the education colleges whose management already runs a degree college (within the distance of 5–10 km) should be allowed to run a four-year integrated B.A./B.Ed. program by sharing the resources of their sister institutions. So, for stand-alone colleges, it would be easier to convert into the composite college if they take one unit of integrated course in social sciences, like B.A./B.Ed. as fewer resources would be required for the same. The other stand-alone institutions without a sister concern in the form of degree college could collaborate with the degree colleges (within the distance of 5–10 km) in their vicinity. For this MOU could be signed between the managements of both the institutions with detailed instructions/modalities/formalities from the policy framer's side or the related institutions.

It should not be just a mass exercise to allot teacher education courses on a mass scale to any and every composite college, as that could potentially cause further deterioration of this prized professional course. Mere addition of a department of education should not be the only criterion for that.

Major curriculum revision workshops are the need of the hour. In the integrated courses, one minor subject can be substituted with pedagogy and the weight of pedagogy gradually increased in the third and fourth years of the program. A full trimester of teaching internship is required. Thus, a teacher education program will comprise of one major, one minor, and pedagogy including the teacher internship. In this structure, the student will qualify the credit of one major to pursue his post-graduation, whereas, for pursuing other courses, his deficit credit may be assessed.

Short-term diplomas and certificate courses would also be made available after completion of the B.Ed. degree for offering more specialized and advanced courses in leadership, management, guidance and counseling for their professional development, as well as career advancement from primary to secondary to senior secondary stages. There must be clear-cut distinction between the diverse education programs, like less than six months, six months, one-year, two-year B.Ed./teacher training programs and four-year B.Ed. programs as well as the open and distance learning (ODL) model of teacher education. Guidelines and equivalence issues should be provided in measurable inputs and outputs so that implementation of these variety of courses does not lead to disputes and ambiguities.

10.4 MEETING THE GLOBAL STANDARDS OF EDUCATION THROUGH NEP-2020

The National Educational Policy 2020 has been drafted in such a way so as to include the recommendations of previous policy frameworks available in this area, such as the Kothari Commission, NCF 2005, The National Policy on Education (1986 and 1992), and so on. It has also included various documents from NCERT and CBSE on school education from time to time. The resultant policy that came out in 2020 is acceptable to most of the stakeholders.

NEP-2020 is accepted internationally by The International Baccalaureate Schools (IB) and International General Certificate of Secondary Education (IGCSE) as a document that meets the global standards of education. This has been indeed a boon for various schools affiliated with CBSE, as the policy is accepted by IB and IGCSE. Similarly, the state boards of education are also accepting it with some modifications suited to their regional needs in order to have local acceptability. The policy strongly recommends teaching in local and regional languages and also in the mother tongue until class five.

NPE-2020 brings relief from traditional styles of teaching and for those who wanted a total transformation of our school system to be on par with the prevailing systems in developed countries. Many of the schools in western countries have given bagless days to their children in primary education. The workload is the bare minimum for the kids, and they feel happy to attend school. Rote memorization has been reduced to a minimum, and more emphasis is placed on play and the Montessori method of teaching. The students are involved in a variety of 'fun and learn' activities. The policy envisions that in time, our children would fair well in Program for International Student Assessment (PISA) where till now we stand last in the list and continue to question the credibility of this exam for students. Similarly, instead of finding fault with the PISA or Annual Status of Education Report (ASER), or for that matter our own system, Performance Assessment, Review and Analysis of Knowledge for Holistic Development (PARAKH), we are obliged to evaluate our mistakes and decide what is to be done to bring our children to the international front.

10.5 RECRUITMENT AND EMPLOYMENT OF TEACHERS

One knows very well about the deteriorated employment conditions of teachers presently. There are still some institutions which are hiring teachers on the caste basis even after 75 years of independence. With the introduction of early childhood care and education

(ECCE), due to NEP, many more teachers will be produced in the future. A point to remember is the quality that we expect the teachers to bring into the school system. Be it any subject teacher, all are required to possess adequate orientation in teaching using novel methodologies following an interdisciplinary and multidisciplinary approach. They are also required to link their teaching to the needs of community and also impart value-based education to their students. The recruitment process must include the assessment of such skills apart from functional domain knowledge. A screening on the basis of some test like the Teacher Eligibility Test (TET) along with an interview should serve as a basis of hiring them to teach in our schools.

Apart from giving a demonstration, clearing the interview, certification of qualifying of local language, and some specialization in IT, teachers must pass the TET after the completion of a B.Ed. degree to gain employment in various government or private schools. The TET will cover all stages primary, secondary, and senior secondary. The NEP-2020 recommends sharing of teachers across a group of schools to meet the shortage of teachers in specialized vocational subjects (music, art, dance, craft, IT) and vocational counselors, social workers, and physical education instructors.

The NEP-2020 also promotes hiring of eminent educationists as 'experts' or 'master instructors' in various subjects, like arts and crafts, horticulture, entrepreneurship, agriculture, etc., to address the problem of shortage of teachers for teaching various vocational and skill-oriented subjects.

Another issue that needs our immediate attention is the meager salaries being paid to the teachers in schools. While the teacher salary structure is proper as per the standard norms in government schools, private schools are free to pay teachers as they see fit, which is generally much less. Moreover, the school lecturers (PGTs) who teach senior classes like +1 or +2, are paid higher salaries as compared to trained graduate teachers (TGTs). The worst to suffer are the primary school teachers (PRTs), who not only have to work really hard to teach a nascent population but also have to adopt kindergarten and Montessori methods to teach the young children. Additionally, they have to teach all subjects to a bigger strength of students at the primary level compared to secondary and senior secondary levels. This differentiation in the salary system exists only in school education. At higher education levels, the assistant professors are paid the same salary, whether they teach the first semester or the final semester. If there has to be a differentiation in payment norms, then it has to be on the basis of qualifications, merit, and experience. The teachers who possess additional skills, and obtain certification by attending professional development programs, may be accordingly considered for salary hikes.

NEP also recommends that such TEIs where Nursery Teachers' Training (NTT), Elentary Teachers' Training (ETT), Junior Basic Training (JBT), and other teacher training programs are being run, they will only be allowed to start B.Ed. and M.Ed. courses provided that they fulfill all infrastructural and other conditions as laid down in the policy frameworks. These programs, if implemented, should be able to achieve the objectives of NEP-2020. The new National Educational Policy has come in 2020 after a gap of 34 long years. It is imperative that the present faculty be trained to meet the outcomes of previous policies, and they need to first unlearn, then relearn the guidelines of the present policy and refresh themselves. It would require the existing TEIs train their teacher educators for the next few years to reorient their teachers to cope with the changes. A variety of seminars, conferences, workshops, and panel discussions may be organized by various bodies for the same.

There is a system of promotion of in-service pre-primary, primary JBT, NTT, or ETT teachers to secondary level, provided they complete their B.Ed. degree, who are not sufficiently trained. Such teachers, when they enter the formal school education system, degrade it considerably. Such promotions have to made on the basis of performance appraisal and not automatic promotion.

Research provides evidence that many of the institutions running JBT, NTT, or ETT do not meet the appropriate standards and lack training in play way, hands-on-and joyful learning activities. The teacher education institutions running JBT, NTT, and ETT programs merely focus on textbook-based knowledge – not on activities which are necessary for fulfilling the developmental needs of children. These Anganwadi workers and preschool teachers are generally given priority to become ECCE teachers, ultimately absorbing them into the existing system of formal education, which is nothing less than a disaster. There are many programs where, we hear, class 10 students may be eligible to serve as ECCE teachers. This is really disheartening. How can a teacher with no proper schooling themselves be made a teacher? The very idea lacks total sensitivity and concern toward the teaching profession.

It is equally important that teachers are recruited having appropriate teaching competencies, IT skills, and a general acumen and interest in teaching rather than merely accommodating teachers who are already employed in this sector, who may be lacking basic competencies. If we are to believe what the majority of the psychologists say that the foundation of future life of children is laid before 5–6 years of age, then how can our nation thrive in the hands of poorly trained primary teachers?

NEP-2020 is visionary in the sense that it mentions 'outstanding teacher' six times in para 5.1–5.7. The country desperately needs good quality and passionate teachers if we are to stride on the path of success. Thus, training to enhance skills, improving job satisfaction, and respecting the status of teachers has been of prime concern in NEP-2020. It is ironic that the school teachers have been frequently utilized for doing odd jobs, such as collecting census data, COVID surveys, pulse polio drives, and serving in municipal, state, or assembly elections, etc. The service of teaching suffers greatly by overworking teachers in these way. Recently, many of the teachers have been found to be COVID-positive performing admission and survey duties, and many have lost their precious lives as well. It is only because they are scared to lose their jobs, while many of the officials working in government offices have been sitting in their air-conditioned rooms. The teaching fraternity has literally lost its dignity, self-respect, and has become a puppet in the hands of bureaucracy. NEP has provisions in para 5.1–5.7 and para 15.1–15.11 regarding the policy support in this matter.

Although there is a general acceptance of the policy decisions, due to its non-controversial nature, there is a feeling among many stakeholders that the policy will soon be forgotten. Many believe that we are already following the majority of the suggestions, and there is hardly anything to be done. Having a policy is one thing, its proper implementation is another entirely. How we do it is yet to be seen. The makers of the NEP-2020 apprehend that it may lose its true spirit in the hands of the defenders whose only intention is to bake their own breads, which is unlike the intention of its creators. Early implementation of policy recommendations is key to the success of NEP-2020.

10.6 TEACHING CAREER AND PROFESSIONALISM

The NEP-2020 aims to create National Professional Standards for Teachers (NPST) by the year 2022 for maintaining performance standards for teachers. This will clearly describe the role and competencies required by the teachers at different stages. It will also serve as a guideline to address major issues of the teachers, such as their promotion, salary, perks, increments, nature of appointment, professional development, etc.

NEP-2020 also advocates performance appraisals, or audits, of the teachers at regular intervals, the criteria for which will be formulated. The promotions and salary hikes of the teachers will not be based on the length of tenure of service but on performance appraisal. Further, the school teachers will have to undergo 50 hours of professional development training every year to keep themselves abreast with the latest information by attending seminars, workshops, lectures, or teacher development modules. Even the principals of various schools will have to attend the compulsory trainings for their professional development related to latest pedagogies, evaluation procedures, team-building, school management and constructivism-based learning. In addition, NCERT will study the pedagogical practices prevalent in different countries at the international level and will bring into force into the curriculum at the national level.

NEP-2020 recommends that the faculty of new TEIs should be derived from the school sector only. The qualified and experienced principals and those possessing necessary qualifications to be eligible for becoming teacher educators may be allowed to work on a deputation basis as the faculty for these institutions. The existing teachers in the schools may be allowed to pursue M.Ed./M.A. education as part of their continued professional development program. There should be provision for study leave for doing the same. The benchmarks for entry as a teacher educator need to be chalked out at length, and the teachers possessing the necessary skill sets in IT, digital learning, content mastery, command in language along with essential soft skills may be considered for the same.

10.7 TEACHER EMPOWERMENT THROUGH NEP-2020

Bolin (1989) refers to the empowerment of teachers to participate in the decision-making process of the school. It includes setting the goals and policies of the school and making decisions about the content and methodology of teaching. NEP-2020 envisions that if teachers are involved in the decision-making process and are provided the freedom and control in their work, then it leads to their empowerment, encourages them to work harder, and enhances their commitment to the students.

NEP-2020 provides autonomy to the teachers and recognizes their worth and contributions to reforming pedagogical practices, thereby enhancing the learning outcomes. Teachers can select appropriate pedagogy to ensure socio-emotional health of their pupils leading to their holistic development. Teachers can adopt innovative teaching methods to achieve the learning outcomes set by them. Thus, their efforts will be recognized and will be shared as recommendations.

It is a general observation that teachers remain isolated and are confined to work in their own schools. NEP-2020 encourages 'school complexes' which refers to collaboration among neighboring schools to reduce teacher isolation. This would lead to frequent interaction and exchange of knowledge, experience, and expertise among teachers. This way less privileged schools get benefit, and communities of vibrant teachers can be created. In order to facilitate creation of school complexes, sharing of school infrastructure, basic facilities of health and hygiene, computer labs, Wi-Fi and internet, libraries, resources for games, sports, and recreational facilities is also encouraged.

10.8 NEP-2020: THE KEY TAKEAWAYS

10.8.1 Drastic Changes in the Structure of School Education

NEP-2020 has kept the students in focus by acknowledging the 5 I's: Indian, Internationalism, Impactful, Interactive, and Inclusive, on the principles of equity, quality, and access. It aims to develop contemporary 21st-century skills among the students both at school as well as the higher education levels. The present 10+2 structure of school education will be replaced by 5+3+3+4 structure for the age groups 3–8, 8–11, 11–14, and 14–18 years, respectively, emphasizing 'early childhood care and education'.

At the undergraduate level, NPE-2020 focuses on holistic multidisciplinary education of 3–4 years of duration. In order to store the academic credits earned by a student digitally from recognized higher education institutions (HEIs), an Academic Bank of Credit (ABC) will be maintained, for awarding the degree on the basis of the total credits earned by the student.

10.8.2 Implementation of NEP-2020

The terms of implementation of the policy guidelines are clearly mentioned in para 27.1–27.3 of Chapter 27, which discusses its implementation in true spirit in a phased manner. The implementation strategy of the policy will be chalked out in consultation with suggestions from all states and UTs. It will be a joint exercise and will be carried out by taking into account the inputs from various stakeholders such as NCERT, SCERT, DIETs, BIETs, CIETs, and prominent educationists and experts in the field.

NCTE launched a portal 'My NEP-2020' to invite suggestions from school teachers and principals as well as from colleges of education across the nation. Various WhatsApp and Telegram groups along with various committees have been formed to take prompt action in this regard.

10.8.3 Impact of NEP Post-Implementation

NEP-2020 emphasizes 'job creators' rather than 'job seekers'. It means that we have to transform our education system to enable the students to think critically, be innovative, and train them in problem-solving skills to produce novel material. The expected pedagogy has to revolve around making education learner-centered, experiential, holistic, heuristic, logical, and, most importantly, joyful. The new curriculum proposed by NEP-2020 will not only include traditional subjects like science, social science, and maths but would also infuse components from arts, literature, sports, music, crafts, culture, and values as per the abilities of the students to develop their overall personality. We have to make education a more enjoyable and dynamic experience for our students. NEP also stresses character-building, enabling learners to be responsible, principled, balanced and empathetic, at the same time preparing them for useful jobs.

NEP also lays stress on bridging the gap between the expected learning outcomes and students' achievement and to have a more integrated, balanced system right from preschool through higher education. The policy strives to achieve the milestones of world-class equitable education for all children in India by the year 2040, regardless of their socioeconomic status.

10.8.4 NEP-2020 and Reforms in Teacher Education

Teacher education is one area that will be overhauled with the implementation of the new education policy. India is a country where a teacher or 'guru' is held in high esteem, who is regarded as even superior to God. Teacher education creates school teachers who shape the destiny of the nation, our students. Thus, such teachers will be required who are equipped with 21st-century skills and, at the same time, are value-oriented. The task of teacher preparation requires multidisciplinary knowledge and multiple perspectives. Thus, NEP-2020 has proposed a four-year integrated B.A./B.Sc./B.Com./B.Ed. degree to be offered only in multidisciplinary composite institutions. NCTE will formulate NCFTE (2021) with NCERT which will focus on a wide range of content and pedagogies along with a rich and varied practicum.

10.8.5 Issue of Vacant Teacher Posts

There are many posts for teachers that have lain vacant for a number of years. NEP-2020 recommends that these will have to be filled at the earliest in a set time frame, especially in rural and disadvantaged areas with high illiteracy rates. The recruitment process will be made more transparent and trustworthy. Local teachers familiar in local languages would be given preference. The in-service teachers would be required to work on their continuous professional development.

The selection would be made on the basis of TET conducted by CBSE along with tests to check content soundness to be conducted by standardized professional agencies along with personal interviews. The process may have little variation as per the requirements of the state. The TEIs would be required to produce competent teachers in different subjects as per the requirements of the states or UTs. This entire process will ensure timely filling up of various posts. The in-service teachers will be granted career incentives and promotions on the basis of their continual assessment of performance.

10.8.6 NEP-2020 and Funding of Education

NEP-2020 is ambitious as far as the funding is concerned in the education sector. It proposes to increase expenditure on education up to 6% of GDP. This expenditure will be made on enhancing the quality of institutions, be it school, college, universities or TEIs. The funds would be provided as per the needs of the sector. The various sources of funding will include funding from center and state government, mechanism of loans for institutions from competent government authorities, loans from philanthropic societies, such as the corporate sector, NGOs, alumnae members, etc. The provision of loans for students with low interest rates, grants by the government being provided by Rathriya Uchchtar Shiksha Abhiyan (RUSA) for enhancing infrastructure and organizing capacity building activities, etc. in HEIs. The government will evolve a mechanism to ensure that the funds are utilized optimally for the intended purposes and in a time-bound manner.

10.8.7 Provision of Merit-Based Scholarships in NEP-2020

NEP ensures to provide merit-based scholarships to the bright students from rural areas wishing to pursue teaching as a profession for the four-year integrated B.Ed. program. There will also be provision of employment opportunities, especially for women, in various schools after the completion of their degree. Special incentives would be provided to those desirous of working in rural areas.

10.8.8 NEP-2020 and Globalization of Education

NEP-2020 attempts to establish global citizens keeping the spirit of 'think global, act local'. NEP recognizes that children learn difficult concepts more easily in their mother tongue. Thus, NEP encourages use of mother tongue or home language as the medium of instruction until at least class 5 or maybe class 8 and even beyond.

The policy provides flexibility in a three-language formula, presently being adopted by various states. States will no longer impose any language formula for the children. This will be entirely the choice of children which language he/she wishes to study, provided that at least two languages are Indian. The policy is open to all languages and does not undermine the importance of any foreign language, say English; rather it promotes multilingualism for greater benefit to the young children.

10.8.9 NEP-2020 and Digital Learning for 'All'

During the COVID-19 pandemic there was a strong need to switch to technology-based learning from face-to-face learning. Initially, there was ignorance, lack of knowledge, hesitation, and resistance among the teachers and students for the technology-centered learning, but with passage of time, as we acquainted and prepared ourselves to the new challenge, there came ample of opportunities to enhance our existing knowledge and receive training that was never required earlier. Virtual learning has offered many learning opportunities and also posed challenges to the teachers as well as students through online learning (Padhi and chiani, 2020). There were many issues that were faced by the teachers across India to cope with this transformation. The biggest challenge that was perceived was the inequality in digital access among people in different strata of society (Dalberg, 2013). The children in rural and backward areas, where there was poor network and internet access, found it a big challenge in their way. A study of NCERT (2020) reflects that approximately

27% of the students do not have access to smartphones/laptops to attend classes in a virtual world. Further, half of the teachers faced problems related to expensive data and slow internet and almost 84% teachers reported facing challenges in delivering the virtual mode of education. The internet usage has also been found to greatly vary among different socio-economic groups in India. This disparity in access to digital technology has greatly disrupted the targets of online education.

Providing and enabling additional economic support for availing the ICTs to unprivileged and marginalized students is the need at this hour (Eamon, 2004). In this regard, a proper institutional mechanism is needed to monitor, assess, and bridge this digital divide, thus taking this pandemic crisis as an opportunity to review the existing disparities in access and uses of ICT, especially among young learners to help achieve equity in learning (Karakara and Osabouhien, 2019).

NEP-2020 has provisions for several initiatives by the government under 'Aatmnirbhar Abhiyan' to bridge the digital divide across the country such as PRAGYATA Guidelines, NISHTHA-Online, Digital Education-India Report, etc. The report of 'PRAGYATA: Guidelines for Digital Education' by the government of India has issued guidelines for limiting the screentime for students of all age groups for digital learning. Further, in the report by KPMG (2020), University Grants Commission (UGC) has also issued specific guidelines for the Indian HEIs it has provisioned through various digital platforms.

These initiatives mean to assure the continuity in school education for 'all' children irrespective of caste, creed, socioeconomic, and cultural background. NETF (National Education Technology Forum) has been formed with an aim to promote frequent use of technology in imparting education. The policy suggests by bridging this digital divide, a wide range of student population will be catered for various educational programs in different languages around the clock. Digital content will have to be prepared in native languages to reach the maximum number of learners. Various existing e-learning platforms, like Swayam, Swayam Prabha, E-Pathshala, NROER, and Diksha offering MOOCs (massive open online courses) will serve as virtual classrooms to have a widespread effect for providing self-learning experiences to them.

10.8.10 NEP-2020 and Some Concerns

NEP-2020 is a 60-page report, and is not voluminous as compared to the reports of early education commissions, policies, and documents. It is like putting the ocean in a pot. It is a reader's delight as it does not take much of your time to read. Many critics claim, however, that it is extremely restricted and is biased toward exclusive schools having elite populations.

Another issue that we have come across is the language controversy, that has demanded our attention for a long time, which now has been appropriately tackled with the NEP-2020. In the past, many eyebrows have been raised as far as the medium of instruction and teaching of languages in schools is concerned. The policy does not impose any 'three-language formula', and offers plenty of choices to the students for studying any local, regional, or even foreign languages. The policy focuses on mother tongue and use of local languages and even use of tribal languages in schools.

There is huge segment of educationists who believe that the policy, if implemented in its true spirit, would certainly be useful for the improvement of school education. All is needed is timely implementation of the policy by the regulatory agencies like NCTE, CBSE, and NCERT. Only time will tell how effective the policy recommendations will be and how many reforms will see the light of day. We, the educationists certainly don't want the policy implementation to suffer in the hands of bureaucrats due to delays in the decision-making process.

It has been observed in the past that many business people who have no link to education or its concerns have started their own TEIs. NEP-2020 strongly advocates that individuals, private groups, or societies who bear no association with the school system, be prohibited from opening new TEIs. Even the existing TEIs in both government as well as the private sector will have to prove themselves through a rigorous assessment procedure in order to qualify themselves for starting new TEIs.

10.9 CONCLUSIONS

Education is an important concern for everyone in the world irrespective of the profession. In each country there might be some issues related to the prevailing educational system and we normally criticize the existing educational systems and suggest some new ways to improve teaching and learning. We also know that the impact of any educational system depends upon effective teachers, which can be produced by quality professional teacher educational institutions. Teamwork is required at various levels, viz. university and institutions, in order to transform the proposed teacher education system. More brainstorming sessions are required in the form of seminars, webinars, and workshops at the institution, university, and the government levels in order to draw out various challenges in the implementation of the policy and also look for solutions to those challenges. The teaching fraternity must be prepared to reflect on the NCFTE 2021 document as soon as it gets released and the apex bodies of teacher education must take a lead role in it to be followed by all other institutions, as perhaps this will set the pace of teacher education programs in the near future.

The 21st-century developmental goals that seek to achieve universal primary education, improved secondary and higher education can only see the light of the day if we revamp the present system of teacher education by extending opportunities to many quality teacher education institutions accredited by NAAC, having good infrastructure and human resources. Many such stand-alone TEIs are perceiving NEP-2020 as a threat to their existence. This would also save a lot in terms of national economy, along with saving the existence of quality institutions by giving them a chance to improve and change with the present times. The dedicated women's education institutions that promote gender equality and work for the cause of empowerment and emancipation of women should be given special consideration in this direction, if we wish to align with the slogan of 'Beti Padhao, Beti Bachao' ('Save the girl child. Educate the girl child'. All stakeholders must work closely with the vision of the policy document for the betterment of the education system of our nation.

As the famous quote by William Arthur Ward goes, 'The mediocre teacher tells, the good teacher explains, the superior teacher demonstrates, and the great teacher inspires', we need teachers who can motivate and inspire their students by their thoughts and actions. Thus, we need good quality teachers, if we wish to see our system of teacher education flourish. NEP-2020 has come as a fresh wave of reforms. The time has come to bring radical reform in the long-decaying system of teacher education.

The four-year integrated stage-specific, subject-specific B.A./B.Sc./B.Com./B.Ed. degree recommended by NEP 2020 to be offered only at multidisciplinary institutions is a way to look forward. By 2030, this will be the minimum degree qualification for teaching in schools, focusing on wide range of content, pedagogy, and practicum training. This will result in only premium quality TEIs to survive; the rest, those which are not able to maintain the quality, will perish. Proper clear-cut guidelines must be provided by the national bodies, like NCTE, for implementing the NEP-2020 in its true spirit. Universities must frame guidelines which can ensure quality for the integrated teacher education programs. It is hoped that NEP-2020, along with NCTE, NCFTE, and NCERT will bring a new and comprehensive curriculum for teacher education, leading to a much-improved system of teacher education.

REFERENCES

Bolin, F.S. (1989). "Empowering Leadership." *Teachers College Record*, 19(1), 81–96.Council for Teacher Education Foundation (CTEF) Punjab and Chandigarh Chapter, http://ctepbchd.com/

Dalberg (2013). "Impact of the Internet in Africa, Establishing Conditions for Success and Catalysing Inclusive Growth in Ghana, Kenya, Nigeria and Senegal." As retrieved from https://www.scribd.com/document/149720583/Dalberg-Impact-of-Internet-Africa-Full-Report-April2013-vENG-FinalAs on 12.8.2021.

Eamon, M.K. (2004). "Digital Divide in Computer Access and Use Between Poor and Non-Poor Youth." *Journal of Sociology and Social Welfare*, 31(2), Article 6.

Karakara, A.A.W. and Osabouhien, E.S. (2019). "Households' ICT access and educational vulnerability of children in Ghana." *Cogent Social Science*, 5(1). https://doi.org/10.1080/23311886.2019.1701877.

KPMG (2020). "Higher Education in India and COVID-19, Impact on Admissions." As retrieved from https://assets.kpmg/content/dam/kpmg/in/pdf/2020/04/higher-education-in-india-and-covid-19-impact-on-admissions.pdf on 15.8.2021.

National Education Policy (2020). Ministry of Human Resource Development. Govt. of India. https://www.education.gov.in/sites/upload_files/mhrd/files/NEP_Final_English_0.pdf. Retrieved on 12.5.2021.

NCERT (2020). "Students' Learning Enhancement Guidelines." Ministry of Education, Government of India. As retrieved from https://ncert.nic.in/pdf/announcement/Learning_%20Enhancement_Guidelines.pdf on 19th Nov. 2020.

Padhi, B. and Chiani, L. (2020). COVID-19 and the Challenges of Virtual Mode of Education in India, CEFT, Xavier University, Bhubaneswar, Odisha. As retrieved from https://practiceconnect.azimpremjiuniversity.edu.in/covid-19-and-the-challenges-of-virtual-mode-of-education-in-india/ *on* 11th Nov. 2020.

Pragyata Guidelines, Department of School Education & Literacy, MHRD (2020). "Guidelines on Digital/ Online Education Provide a Roadmap for Carrying Forward Online Education with Enhanced Quality." As retrieved from https://iasbhai.com/pragyata-guidelines-upsc/ on 20.8.2021.

WEB REFERENCES

https://devsamaj.org/two-day-national-webinar-on-insights-of-national-education-policy-2020-higher-school-education.php dt. Retrieved on 31.8.2020–1.9.2020.

https://devsamaj.org/national-education-policy-2020-webinar.php dt. Retrieved on 29.12.2020.

http://www.devsamaj.org/workshop-on-new-education-policy-2020.php dt. Retrieved on 18.2.2021.

https://timesofindia.indiatimes.com/blogs/edutrends-india/nep-2020-empowering-the-teacher/. Retrieved on 20.5.2021.

https://www.ugc.ac.in/pdfnews/5294663_Salient-Featuresofnep-Eng-merged.pdf. Retrieved on 21.5.2021.

CHAPTER 11

Techno-Pedagogical Concepts with Digital Skills

Divya Nair, Neeta Mhavan, and Asawari Bhave Gudipudi

CONTENTS

11.1 UNDERSTANDING PEDAGOGY

Jhanvi has opted for two electives in her degree course. Ms. A, a highly experienced teacher, is the instructor for elective one, and Ms. B, a comparatively less experienced teacher, is the instructor for elective two. Although Ms. A's content knowledge is exceptional, Jhanvi experiences difficulty in comprehending the simplest of concepts. She keeps enduring this difficulty in every class of Ms. A. Jhanvi finds Ms. A's explanations quite monotonous. Raising questions in order to get clarity on the concepts only adds to the complications of

DOI: 10.1201/9781003254942-11

the problem, as Ms. A fails in delivering the concepts in a lucid way. Jhanvi has started to dislike the sessions of Ms. A. She also finds the initially simple subject to be complicated after attending Ms. A's sessions. The reverse case is seen in Ms. B's sessions. There is novelty in every concept that is delivered in Ms. B's class. Jhanvi noticed that as per the requirement of the content, the teaching methodology employed by Ms. B differs. She awaits excitedly for her elective two classes and doesn't miss any of Ms. B's sessions, even though content-wise the difficulty level of the second elective is comparatively higher than elective one.

This type of an experience that Jhanvi is undergoing might be familiar to many of us, especially as students. Beyond the shadow of a doubt, strong content knowledge is a pre-requisite of a good teacher. However, there are some additional requirements that will raise the bar of a teacher from being good to outstanding and admirable. One of the most important necessities that will decide the success of your teaching and learning journey is the pedagogy that you choose as a teacher.

Pedagogy is derived from the Greek word 'paidagogiko', which means 'leading the child' (Shah, 2021). This typical etymological meaning of this term is irrelevant in the 21st century classroom. Learners are no longer passive spectators but are expected to be active contributors in their own learning. Constructivism and active learning have been emphasized in all important educational policies be it National Curriculum Framework (NCF) 2005 or National Educational Policy (NEP) 2020. Thus, we cannot stick with the primitive concept of pedagogy as a process of leading the learner nor can we simply attribute it superficially as a process of teaching. Hence, pedagogy needs to be defined comprehensively.

For purposes of clarity, let's define pedagogy as the process that bridges the gap between teaching and learning by utilizing appropriate and relevant methods, practices, activities, techniques, and strategies with the objective of making the entire learning process meaningful. Additionally, it is the means for motivating students to take a lead in constructing their own knowledge for practical implementation in dealing with different, novel, and challenging life experiences during their lifelong journey as learners.

Now that we have some idea of what pedagogy is, it will be easy to interpret Jhanvi's difficulty during Ms. A's lectures and her budding interest in Ms. B's sessions. Even though Ms. A is highly experienced and has a strong hold on her content, she is unable to put it across to the students in an interesting way, as she is not utilizing a well thought out and relatable pedagogy. Thus, in her case, the teaching–learning experience becomes totally ineffective. Ms. B is putting the pedagogy into practice by using innovative and objective-based needs. Thus, even though she lacks experience, her way of involving the students and giving meaning to the entire learning experience is extremely practical as well as constructive.

With the above hypothetical example, it is clear that merely understanding the concept of pedagogy is not enough. Equal emphasis should be given to a teacher's role in the success or failure of any pedagogical approach. Rigidity and failure to blend as per the dynamic changes that take place in the educational setup can hamper the pedagogy and its ultimate attainment. Interpreting the need of current times and designing appropriate pedagogical considerations should also be given prime importance. The Covid-19 pandemic has compelled every educator to shift into an online mode of education. For a few of the educators, this is a welcome change, but for the remaining, online education is a forced transformation that they were loathe to comply with. It is important for educators to understand that teaching through technology isn't confined to video conferencing applications like Zoom, Microsoft Teams, Google Meet, etc. There are different facets that every educator should explore while integrating technology into their pedagogical considerations. This thought opens the door for discussion of the concept, 'technologically enabled pedagogy'.

11.2 WHAT IS TECHNOLOGICALLY ENABLED PEDAGOGY?

After undertaking a training on basic formatting tasks related to MS-Word, Anup decided to acclimatize his students with his new knowledge. In one of the sessions, Anup confidently started introducing formatting aspects to his students. Before he could go ahead with his demonstration, the students of his class started suggesting an alternative way of formatting by using the LaTeX program. According to the students, LaTeX was a better and more flexible choice compared to MS-Word. This suggestion of the students made Anup highly uncomfortable as he had no idea of what the students were talking about. How should Anup handle this situation?

Even in the simplest of situations there is a constant bombardment of new information that enhances the knowledge tapestry magnificently. Recent years have seen many major breakthroughs in the implementation of technology, particularly in the education sector. Thus, there is a constant need to keep ourselves updated and alert. Anup's objective of attending the training program was to bring about a refinement in his technological expertise. This was an excellent initiative from his side as a teacher. The dilemma that he is facing in the class is due to his failure of judging the potential of his technologically native students and not because of his resistance to embracing new technological advancements.

There are two ways through which Anup can deal with this situation. First, by ignoring the student's suggestions and continuing with what he knows best. This will, however, be an extremely rigid way of dealing with the problem. The second way out from this situation is by exploring the suggestions of the students. This will boost the enthusiastic streak of the students, as they will feel involved in the teaching–learning process. Moreover, this strategy also highlights one important point, namely, that the teacher need not know how to use every single technological tool, but they should be flexible in accepting and learning the technological dimensions if the situation arrives. This forms the crux of technologically enabled pedagogy.

Technologically enabled pedagogy can be understood as a positive outlook toward newly emerging digital strategies and approaches through which the students get involved and engaged in active and independent learning tasks by inculcation of digitally competent skills and information that will enable them to be contributing members of the technologized workforce.

Thus, giving laptops to the students or providing them with Wi-Fi connectivity is not what defines technologically enabled pedagogy. Empowering the students with digital hand-holding as well as refining your own teaching approaches as per the objectives and expected learning outcomes will lead to the successful amalgamation of technology and pedagogy. Using a wide diversity of technological tools is not important. Rather, the focus should be on using the correct technological medium at the right time and in the correct way.

11.3 SIGNIFICANCE OF TECHNOLOGICALLY ENABLED PEDAGOGY

Human beings are good at adapting. However, getting oneself reformed as per the wave of technologically enabled pedagogy is not just for formality's sake. Becoming competent in technologically enabled pedagogy has its own important reasons (Figure 11.1).

Let's travel back a few decades to the period wherein there was no trace of the internet. Sources of information were restricted to the morning newspapers or the prime-time news bulletins. There were no mobiles, and people had to depend on postal services for delivering messages to their loved ones through letters or telegrams. Life was definitely meaningful but less complicated compared to the present scenario. With the explosion of technology, every lexicon of life has become connected globally. The entire information domain can be opened with a single click of one's finger. The power of technology has also given rise to a plethora of challenges. The skill set that was prevalent in the good old era is no longer considered useful in the digitally equipped world. Thus, both teachers and students should master the elements of technologically enabled pedagogy.

A single look at your hand will make it quite evident that none of your fingers are alike. This example can be a good analogy for relating to the diverse needs of learners. To put it in a straightforward statement, none of our learners are identical and, therefore, their learning needs should also be catered to in a differentiated way. Technologically enabled pedagogy can enhance the experiences given to students by taking into consideration their individual differences. For instance, by using asynchronous self-paced projects, the needs of those students who find it difficult to grasp the concept in one go can be gratified. Similarly, not everyone who is a part of this era is technologically competent. Systematic training can be given to the students for improving their digital skill set by personalizing the lessons with the help of technologically enabled pedagogy.

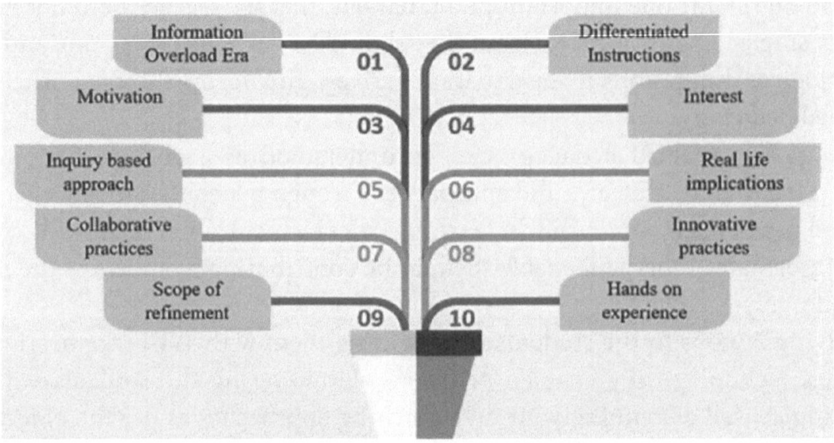

FIGURE 11.1 The significance of technologically enabled pedagogy.

Consider this. You are busy doing your routine tasks, when a message gets delivered in your inbox. The HR team has informed all its employees about the annual appraisal drive. There is a comprehensive form that needs to be filled along with relevant evidences. As you go through the form, you notice that a huge weight has been allotted to digital skills and professional development activities of the faculty. You are highly disappointed with yourself, as you have nothing to upload in any of the mentioned norms. You immediately start searching for some professional development course in order to fill the gap so as to stay in the race of the annual increment. This is an example of a well-known concept called motivation. In simple words, motivation is the drive that shapes our behavior. It is goal-oriented and need specific. The process of learning can be immensely successful if the learners are motivated and if the learning caters to their areas of interest (Barberan, 2020). Today's generation loves to get involved in technology-related tasks. The element of intrinsic motivation is present by default in learners as they not only live with technology but also breathe in technology. Thus, using a technology-based pedagogy will motivate the learners to get actively involved in their learning.

Technologically enabled pedagogy also provides the scope of the inquiry-based learning approach. For instance, a task of referencing may appear simple on paper. However, as the student starts working on the project, he or she will realize that picking up any random source from the internet is not reliable nor valid. The authenticity of the source and the related content should be mapped. Successful completion of the referencing task will require the student to follow an inquiry-based approach and a systematic problem-solving path with the aid of technology.

Technological competence is extremely important in all our day-to-day life experiences. In 2016, the overnight demonetization announcement in India boosted the use of cashless transactions. Without an ounce of doubt, online transactions were definitely in existence before the demonetization drive. However, a part of the population was undeniably cold toward availing the benefits of cashless transactions before this revolution was forcefully imposed on one and all. This is a classic example that technologically enabled pedagogy is not restricted to the classroom cum laboratory setup, but it also has wider implications in our real-life problems and experiences.

Another important feature of technologically enabled pedagogy is the collaborative and innovative outlook. Collaborative approaches have been in the picture in offline settings for many years. However, the teacher should design her lessons in a way through which the students actively collaborate in interesting and innovative projects. Some of the simplest yet effective ways through which technologically enabled pedagogy can promote collaboration of its learners is through facilitating discussions via an online forum, blogs, encouraging the use of Google docs, declaring assignments via Insert Learning, making smart use of social media for learning purposes, etc.

Coming back to one of the points that had been raised while discussing the definition of technologically enabled pedagogy, it is crucial to appreciate the fact that none of us will be knowing every aspect of technology in a full-fledged way. However, this pedagogy provides a scope of refinement with respect to our skills and information set. Since the teachers and students will be dealing with technology in real time, this will give them a practical touch and thus increase the effectiveness of the technologically enabled pedagogy practices. This will further bridge the gap between theoretical elements and hands-on experiences.

11.4 DIGITAL IMMIGRANTS AND DIGITAL NATIVES

After suffering a massive loss in his business, Raghav decided to shift base from India to his best friend's place in France. Though he was good at his craft, Raghav faced a lot of problems in understanding the French language. His discomfort in English also increased his communication woes. He somehow managed to survive in this new country due to the support of his best friend. With the passage of time, he established his business and his family also got settled decently. His daughter Shreya, who grew up in France, could master French with a perfect dialect. However, even after living in France for so many years, Raghav is still apprehensive when it comes to French as well as English.

If we think of this example in terms of technological skills, then Raghav would be a 'digital immigrant', and Shreya would be a 'digital native'. This interesting concept of digital generations was put forth by Mark Prensky.

People who were born in the times when there was technological exposure are called 'digital natives'. These are usually the people who were born after 1980s. They are comfortable using technology as it was already a part of their existence (Autry and Berge, 2011). Comparing it with Shreya's case, she was able to learn French and communicate effectively as she was brought up in France, and communicating in French was a part of her natural world. Digital natives are further categorized as millennials, who were born between 1980 and 2000, and Generation Z, who were born after the year 2000 (Mohr and Mohr, 2017).

Coming back to Raghav's case. For Raghav, this transition was not natural. He was only comfortable in his native language as he had spent most of his life using the same. Though the circumstantial challenges forced him to adapt to new changes, he couldn't completely familiarize himself with the new demands of language acquisition. This is typical of the population called 'digital immigrants'. They are people who were born before the 1980s, wherein technological advancements were not dominant (Prensky, 2001). There is a sense of rigidity in their outlook toward technology, as their exposure toward the same took place in the latter half of their life. Subconsciously there is a slight fear in their minds that makes it difficult for them to be wholeheartedly involved in the technologically enabled environment.

Let us consider another example. How would you feel if you were asked to climb a 15-story building instead of taking the elevator? This irrelevant demand of climbing the stairs when there is a lift can create a conflict for the people involved. Similarly, the inflexible attitude of most digital immigrants can hamper the techno-pedagogical requirements of the digital native learners. A digital immigrant teacher will most probably view things from his perspective, and will try to use strategies that were prevalent in his times as a student. They will be stuck with the slow and systematic processes which will be contradictory to the multitasking super speedy requirements of the students.

For catering to the characteristics of this generation, the digital immigrants need to take a huge leap. This leap should be structured with a flexible mindset, openness to new technological methodologies, and a respectful attitude toward the digital natives' characteristic and dynamic learning traits. Give them tasks that might be challenging for them but also enjoyable. For instance, instead of asking them to download a movie clip, you can ask them to create their own movie using any of the moving-making tools. Similarly, instead of insisting on face-to-face collaborations you can initiate a research project that involves social media dynamics and allows the student to explore a self-led social media research. It is also important to note that while dealing with this generation of learners, you are not a teacher but a facilitator. Technology will never replace you however; you have to provide the correct environment and opportunity for designing a well-equipped digital learning environment for the optimum progress of your learners' abilities. If you are successful in doing this then there won't be any digital gap between the digital immigrants and the digital learners.

11.5 DIGITAL LITERACY

In its simplest definition, literacy refers to the ability of reading and writing. Digital literacy is also a skill set of technologically equipped abilities (Eshet, 2004). For better understanding let us consider a simple example.

Samiksha recently took possession of her flat. She decided to work on the furnishing aspect first. Upon contacting a vendor, she got a quote of 1.5 lakhs that covered just her kitchen furnishing needs. As per the discussion, the entire house furnishing would cost her approximately 7–8 lakhs. The neighboring vendors also quoted a similar price. Samiksha was confident that the price was too high due to the commission of the agents involved in this task. She decided to contact vendors from different areas independently without involving any agents. After a thorough search, Samiksha was able to get the entire furnishing deal for her house in good quality material at the cost of 2.5 lakhs.

If we analyze Samiksha's experience, a few points can be easily highlighted. Samiksha started the process of furnishing her house by searching for different vendors, Technologically this process can be termed 'searching for the information source'. Her initial meetings with the vendors gave a search result that was not appropriate as per Samiksha's needs. Her realization that the quote was an exaggerated one, highlights one of the most important skills of digital literacy, that is, examining the authenticity of the information source. Toward the end, Samiksha was able to get a fantastic deal without compromising on the quality or her needs. This can be related to making use of the correct information obtained through technology without compromising on the validity and authenticity of the information.

Thus, digital literacy can be defined as a set of technologically desired skills which enables the learner to locate the correct source of information, critically validate the authenticity of the acquired information, and make appropriate use of the validated information at the right time at the right place.

These skills need to be learned by every stakeholder who uses technology. There is a primary misconception that whatever information one gets through the search engines of the internet is accurate. Most of the time, people just see the first few hits of the search results and, without checking the source or the appropriateness of the information, they start using this information. As teachers who are devoted to the technologically enabled pedagogy, it is our duty to teach the appropriate referencing skills to our students. Every piece of information available on the net is not necessarily correct. This should be considered as a mantra by all students and teachers if they are to be called digital literates in the true sense.

11.6 DIGITAL PEDAGOGY TOOLS

According to the latest research, integration of technological tools in teaching and learning process can support constructivist and meaningful learning (Sailin & Mahmor, 2018). Similarly, there is enough scientific evidence to demonstrate that activities in which students are involved completely and in different ways can lead to good attention and better retention abilities (LaDage et al., 2018). Actively engaging the students in a digital platform can be a challenging as well as an exciting task. In both the cases, the final outcome can be a big failure if the technologically enabled pedagogy lacks precision. In any digital platform, interaction and participation in the online activities by the learner are extremely important. Thus, the teachers should focus on providing an environment that will inspire the students to manage their own learning practices. The teacher should also be critical about the choice of online activities and digital tools. It is important that the selected activity and tool should serve the purpose and not be used just for formality's sake. Educators should also make maximum utilization of the public digital library of Open Educational Resources (OER) Commons for organizing, creating, and sharing knowledge-based resources over a huge online network in a collective approach (https://www.oercommons.org/). In this section, we will be concentrating on few digital pedagogy tools that can be used efficiently for creating a collaborative and active student-oriented digital learning experience. An attempt has also been made to quote the examples of open educational resources wherever possible.

11.6.1 Interactive Presentations

PowerPoint presentations are quite popular in the education sector. The concept of presentations can get an interesting twist by using Pear Deck (https://www.peardeck.com/googleslides). It can be described as an interactive classroom response system (Mache et al., 2017) that is available as an add-on for Google slides. It works well on almost every digital device. There are in-built lesson builders right from the start of a lesson till the end of it. It allows the user to generate stimulating problems, challenging activities, and one-to-one discussion with the students (Haryani & Ayuningtyas, 2021). Critical thinking, social skills, and subject-specific activities can also be utilized for appropriate individualization of the lessons. Staying true to the vision of online teaching, Pear Deck provides the facility of student-paced activity and instructor-paced activity. Student-paced activity is well suited for an asynchronous way of learning, whereas instructor-paced activity refers to the synchronous learning mode best suited for live sessions. Every educator should explore this interesting tool for removing the monotonous appeal of the commonly used PowerPoint presentations. This will certainly provide an interactive platform that will be both informative as well as enjoyable for the students.

11.6.2 Digital Storytelling

At its core, digital storytelling allows computer users to become creative storytellers through the traditional processes of selecting a topic, conducting some research, writing a script, and developing an interesting story. This material is then combined with various types of multimedia, including computer-based graphics, recorded audio, computer-generated text, video clips, and music so that it can be played on a computer, uploaded to a website, or burned on a DVD (Robin, 2008). It is one of the best tools that can be employed in a flipped classroom setup. Teachers can also make this a group activity by bringing the element of collaboration to an online environment. StoryboardThat (https://www.storyboardthat.com/), Renderforest (https://www.renderforest.com/), and Voicethread (https://voicethread.com/) are some of the interesting digital story-making software that every educator should explore.

11.6.3 Blogs

Blogs can enhance learning through insightful sharing of opinions, information, and thoughts. It can also be used successfully in a flipped classroom setup, in the form of a home activity or post-discussion activity. Writing a blog on any subject-specific topic or on an important issue can be given as an assignment to the students. Blogs can also be used as a medium of critical and meaningful discussions between the students. WordPress (https://wordpress.com/), LifeType (https://www.opensourcecms.com/lifetype/), and Roller (https://rollerweblogger.org/project/) are some of the open-source blogging platforms that can be utilized by the educators (Keith, 2014).

11.6.4 Podcasts

It is a digital audio file comprised of either audio or video MP3/MP4 recordings that can be downloaded directly to the desktop computer as well as to various mobile devices (Harris & Park, 2008). Podcasting can be one of the best ways for catering to the needs of learners who have an auditory learning style. In academics, podcasts can be viewed as administrative podcasts (e.g. general information, guides), special lecture series (guest lectures, commencement lectures, etc.), and classroom podcasts (anything involving curriculum teaching). Some of the open-source platforms that can be used for preparing a podcast are Audacity (https://www.audacityteam.org/) and Juice (http://juicereceiver.sourceforge.net/)

11.6.5 Virtual Laboratories

The virtual laboratory is a popular ICT-enabled learning tool, which is essentially a computer-based activity, where students interact with an experimental setup or conduct other activities through a user interface (Achuthan et al., 2020). Virtual laboratories can be used for active individualized learning, simulation activities, practical experimentations, and interactive probing tasks. One of the best virtual labs that provides the experience of a physical laboratory in an online environment is OLabs (http://www.olabs.edu.in/). A variety of experiments ranging from subjects like physics, chemistry, biology, mathematics, and English is available for exploration at this site.

11.7 ASSESSMENT METHODOLOGIES

Assessment strategies in an online setup should not ideally follow the path of the rudimentary evaluation methodologies of offline teaching. Different innovative and interesting assessment possibilities should be incorporated along with a balanced blend of graded examinations on the basis of the online task assigned to the students. Let us take a look at some of the assessment strategies that can work very well in an online learning environment.

11.7.1 Rubrics

Rubrics are usually used for graded activities. It comprises in-depth details of those pointers on the basis of which the learners will be marked in the said activity. Google Classroom provides an in-built rubric template that can be customized as per the requirements. Google Classroom (https://edu.google.com/intl/en-GB/products/classroom/) also provides the facility of generating an automatic report card once the assessment is complete. Another easily accessible source for getting good rubric references is the iRubric platform which is a comprehensive rubric development, assessment, and sharing tool that supports a variety of applications and is absolutely free to use and share (https://www.rcampus.com/indexrubric.cfm). Rubrics bring transparency into the grading system for both the students as well as the teacher.

11.7.2 Discussion Analysis Tool

A lot of discussions and group activities take place in an online teaching–learning environment. It becomes a herculean task for the teacher to go through each and every line of the discussion and assess the same fairly. One of the ways through which discussions can be assessed is through the use of Discussion Analysis Tool (DAT). It is used to identify patterns in interactions and determine which interactions promote critical thinking (Jeong, 2003).

11.7.3 Game-based Formative Assessment Tools

The basic objective of formative assessment is to monitor the student's progress and to bring about an improvement in his overall performance graph by providing continuous feedback and through appropriate remedial teaching strategies. Game-based online formative assessment tools have become a quick and popular way for getting immediate feedback of the learner's understanding and inclusive advancement. Research supports the fact that assessment tools based on the concept of gamification promote active learning, motivational behaviors, and enhance critical thinking and problem-solving abilities of the learners (Özer & Şad, 2019). Kahoot! (https://kahoot.it/) is a popular gamification tool used for formative assessment purposes. Apart from Kahoot! other examples of gamification platforms that can be used efficiently for assessment purposes include Quizziz (https://

quizizz.com/), Flipgrid (https://info.flipgrid.com/), Quizlet (https://quizlet.com/), and plickers (https://get.plickers.com/).

11.7.4 Integrating Assessment in an Ongoing Class

Assessment can also be a built-in feature of your lessons, irrespective of the mode you opt for, i.e., synchronous or asynchronous. Again, this way of assessment can be a good formative strategy. One of the excellent sources for this facility is InsertLearning (https://insertlearning.com/). It is a Google Chrome extension that can convert any web page into an interactive website. Students and teachers can engage in discussions through an active discussion forum, questions can be posed, and students can put across their answers in the same document. With a sticky note feature, InsertLearning also allows sharing of media, whether images, videos, or a simple link in the same interactive document. Another assessment-based tool that can be used for this purpose is Edpuzzle (https://edpuzzle.com/). It is basically a video-based assessment tool wherein the teacher can embed different types of textual or audio-based questions within the video at different junctions. The students won't be able to proceed ahead until and unless they answer the questions appropriately. Hence, this is a wonderful assessment tool for tracking students' progress in real time (Ware, 2021).

11.8 THE ROLE OF TEACHERS IN A TECHNO-PEDAGOGICAL LEARNING ENVIRONMENT

Teaching and learning in a technologically enabled pedagogy are very much similar to that of an offline educational framework. There will be involvement of content, instructions, activities, and assessments, irrespective of the platform used. However, some new roles need to be filled by both the teachers and students as they transit to the online space. Teachers will no longer be instructors, nor would the learners be passive listeners. Both will have to contribute equally in the technologically enabled learning space. Thus, they will be partners and collaborators right from the commencement of a teaching–learning activity till the culmination of the same. The focus of the teacher and student should not be on information exchange; rather it should be directed in fostering a hub of interaction and collaboration.

Making students independent learners and in-building the aspect of self-motivated behavior within the students will be a huge task for the teacher. This problem can be simplified with appropriate selection of strategies and activities that will set the base of an enthusiastic engagement from the learner's side. While designing the instructional aspects of an online class, the teachers should make use of the strengths of this format. The major advantage of a technologically enabled pedagogy is the support it offers to various formats. If a teacher makes smart use of the various online resources, flexible tools, and available digital formats, then she can successfully construct a lesson based on psychological principles such as multiple intelligences, learning styles, individual differences as well as differentiated instructions to name a few (Subban, 2006). Thus, one of the major roles that a teacher should cater to is consideration of the learner's needs and leaving the techno-pedagogical environment flexible enough for timely modifications and enhancements.

In a technologically enabled environment, the teacher should also fill the roles of a good planner, role model, coach, facilitator, and communicator (Heuer & King, 2004). Preplanning and introductory activities should be planned systematically by the teacher. They should be able to use innovative strategies like blended learning and cooperative techniques, with appropriate modifications in an online setting.

Apart from the above-mentioned points as per research, the teacher's role in an online space can be grouped into pedagogical, social, managerial, and technical aspects. Educational facilitation is taken care by the pedagogical aspect, social role focuses on creating a social space for online learning. The managerial role focuses on agenda setting, pacing, objective setting, rulemaking, and decision-making while the technical role depends on the instructors first becoming comfortable with the technology being used and then being able to transfer that level of comfort to their learners (Keengwe & Kidd, 2010).

In addition to the above-mentioned points, teachers should also make efforts toward molding the students in becoming independent learners and social collaborators. This can be done by using a combination of various strategies. Self-paced activities and independent projects should be encouraged. While working in groups, every learner should be made accountable for his or her contribution toward the successful completion of the task. Variations should be introduced while making use of interactive tools for teaching and assessment along with the implementation of a timely feedback process. This will increase their curiosity level and will keep them motivated for the next set of activities.

11.9 CONCLUSION

There is a huge wave of technological advances within the educational domain. From being considered a powerful teaching aid to becoming one of the indispensable and irreplaceable parts of the teaching–learning directives, technology has certainly come a long way. Like humans, the digital scenario has also been subjected to a variety of changes due to various inventions and researches. Thus, it is now beyond time that we accept technology and adapt it logically as a pedagogical approach for attaining the pinnacle of the teaching and learning process. In a technologically enabled pedagogical practice, teachers will no longer be the torch-bearers of instruction. They have to break the barriers of the conventional teaching approach by unlearning a few rigid practices and embracing the flexible tactics of digital pedagogy (Lewin and Lundie, 2016). Similarly, the students should also put in their best effort toward becoming independent learners as well as social collaborators who can switch their roles effortlessly as needed. Technology will never replace the teachers, but the pedagogical knowledge blended appropriately with the correct element of technological expertise will certainly lead to a progressive and qualitative growth in the domain of education. Thus, it is important to perceive the digital pedagogy as a new and promising revolution that will not only enhance the teaching–learning process but will also weave a path of progress for every visionary stakeholder of education.

REFERENCES

Achuthan, K., Nedungadi, P., Kolil, V., Diwakar, S., & Raman, R. (2020). Innovation Adoption and Diffusion of Virtual Laboratories. *Innovation Adoption and Diffusion of Virtual Laboratories*, *16*(9), 4–25. https://doi.org/10.3991/ijoe.v16i09.11685

Autry, A. J. Jr, & Berge, Z. (2011). Digital Natives and Digital Immigrants: Getting to Know Each Other. *Industrial and Commercial Training*, *43*, 460–466. https://doi.org/10.1108/00197851111171890

Barberan, M. F. Z. (2020). The Motivation and its Importance in the Teaching-Learning Process. *International Research Journal of Management, IT& Social Sciences*, *7*(1), 138–144. https://doi.org/10.21744/irjmis.v7n1.832

Eshet, Y. (2004). Digital Literacy: A Conceptual Framework for Survival Skills in the Digital era. *Journal of Educational Multimedia and Hypermedia*, *13*(1), 93–106. https://www.learntechlib.org/primary/p/4793/

Harris, H., & Park, S. (2008). Educational Usages of Podcasting: Colloquium. *British Journal of Educational Technology*, *39*(3), 548–551. https://doi.org/10.1111/j.1467-8535.2007.00788.x

Haryani, F., & Ayuningtyas, N. (2021). The Impact of Interactive Online Learning by Pear Deck during COVID-19 Pandemic Era. *Journal of Physics: Conference Series*, *1957*(1), 012006. https://doi.org/10.1088/1742-6596/1957/1/012006

Heuer, B., & King, K. (2004). Leading the Band: The Role of the Instructor in Online Learning for Educators. *The Journal of Interactive Learning*, *3*(1), 1–11.

Jeong, A. C. (2003). The Sequential Analysis of Group Interaction and Critical Thinking in Online. *American Journal of Distance Education*, *17*(1), 25–43. https://doi.org/10.1207/S15389286AJDE1701_3

Keengwe, J., & Kidd, T. T. (2010). Towards Best Practices in Online Learning and Teaching in Higher Education. *MERLOT Journal of Online Learning and Teaching*, *6*(2), 533–541.

Keith, A. (2014). Blogging in the Classroom: Using WordPress Blogs with BuddyPress Plugin as a Learning Tool. *Journal of Advertising Education*, *18*(2), 5–17. https://doi.org/10.1177/109804821401800203

LaDage, L. D., Tornello, S. L., Vallejera, J. M., Baker, E. E., Yan, Y., & Chowdhury, A. (2018). Variation in Behavioral Engagement during an Active Learning Activity Leads to Differential Knowledge Gains in College Students. *Advances in Physiology Education*, *42*(1), 99–103. https://doi.org/10.1152/advan.00150.2017

Lewin, D., & Lundie, D. (2016). Philosophies of Digital Pedagogy. *Studies in Philosophy and Education*, *35*(3), 235–240. https://doi.org/10.1007/s11217-016-9514-7

Mache, J., Tan, N., Shoemaker, G., & Weiss, R. (2017). Pear Deck: An Interactive Classroom Response System to Encourage Student Engagement. *Journal of Computing Sciences in Colleges*, *33*(1), 156–158.

Mohr, K., & Mohr, E. (2017). Understanding Generation Z Students to Promote a Contemporary Learning Environment. *Journal on Empowering Teaching Excellence*, *1*(1), 84–94. https://doi.org/10.15142/T3M05T

Özer, N., & Şad, S. N. (2019). Using Kahoot! as a Gamified Formative Assessment Tool: A case study. *International Journal of Academic Research in Education*, *5*(1), 43–57. https://doi.org/10.17985/ijare.645584

Prensky, M. (2001). Digital Natives & Digital Immigrants. *On the Horizon*, *9*(6), 1–6. https://doi.org/10.1108/10748120110424843

Robin, B. R. (2008). Digital Storytelling: A Powerful Technology Tool for the 21st Century Classroom. *Theory Into Practice*, *47*(3), 220–228. https://doi.org/10.1080/00405840802153916

Sailin, S. N., & Mahmor, N. A. (2018). Improving Student Teachers' Digital Pedagogy through Meaningful Learning Activities. *Malaysian Journal of Learning and Instruction*, *15*(2), 143–173. https://doi.org/10.32890/mjli2018.15.2.6

Shah, D. R. K. (2021). Conceptualizing and Defining Pedagogy. *IOSR Journal of Research & Method in Education*, *11*(1), 06–29. https://doi.org/10.9790/7388-1101020629

Subban, P. (2006). Differentiated Instruction: A Research Basis. *International Education Journal*, *7*(7), 935–947. https://eric.ed.gov/?id=EJ854351

Ware, E. (2021). Edpuzzle. *Journal of the Medical Library Association: JMLA*, *109*(2), 349–350. https://doi.org/10.5195/jmla.2021.1202

CHAPTER 12

National Education Policy-2020

A Landmark in India's History of Education

Neeta Mhavan, Divya Nair, and Asawari BhaveGudipudi

CONTENTS

DOI: 10.1201/9781003254942-12

12.1 DEVELOPMENTS IN THE INDIAN EDUCATION SYSTEM

The educational structure and system in India have profound ancient roots from which the contemporary education system has been developed. The ancient education system of India was culturally and spiritually profound. The history of the Indian education system can be traced through four distinct eras categorized as ancient, medieval, colonial, and post-colonial eras.

12.2 THE ANCIENT PERIOD

The ancient period witnessed two major systems of education 'the Vedic', and 'the Buddhist'. The ancient education system stressed that education should prepare individuals to achieve the aim of emancipation, that is to be free from all the restrictions of life. Religion was the foundation of ancient education. As religion played predominant role self-realization or 'Moksha' was the eventual aim of education accompanied by preservation and transmission of ancient Indian culture. The temples and the community centers performed the role of schools.

12.2.1 The Vedic Education System

The education during the Vedic period was the first systematic teacher-taught schooling that evolved in ancient India. This education system was very developed and student-centered, and the primary aim was the infusion of skills and values. Education was infused with due proportion to unfurl the skill sets and moral potential of the individual. The education was richly embedded in the culture, moral values, and ethos of India.

The foremost intent of Vedic education was the all-around development of the pupils to lead a harmonious life with nature and society. While doing so, this system laid emphasis on intellectual, physical, moral, and spiritual abilities. In ancient India, both formal and informal modes of education thrived. Indigenous education was imparted at home and at the gurukuls. Knowledgeable people who were learned and experts from society steered the young learners in imbibing the wisdom to lead a balanced life. In those times, temples were the centers for learning and places to promote education. It was that time that people from all parts of the world like China, Greece, etc., visited India, as India was the only place where education flourished as even at that time there were universities to gain higher knowledge. Teaching was largely based on guru-shishyaparampara where the students had to memorize intricacies of grammar, logic, and philosophy and had to reflect upon what was taught in the class. Women's education was liberal during the Vedic period. The Vedas gave honorable and respectable status to women. Girls had equal rights to education with boys. Thus, during the Vedic period, education was quite evolved from language to science – which included the study of philosophy, logic, metallurgy, medicine, history, architecture, astronomy, polity, agriculture, husbandry, commerce, trade, etc.

12.2.2 The Buddhist Education System

In India, the Buddhist period started approximately in 600 B.C. and continued to exist until 600 A.D. Buddhist education practices were more inclined toward religious connotations.

The most important aspect of Buddhist education practices was access to all, irrespective of caste, religion, or class. The major objective of this system was to help individuals attain 'Nirvana'; in other words, attain a transcendent state which is free from suffering, desire, or sense of self, through knowledge, wisdom, and meditation. Buddhist education believed that education should enable individuals to free themselves from the cycle of death and rebirth. It also argues that the ultimate enlightenment or 'Anuttara-Samyak-Sambhodi', is not something that can be acquired but is inherent and individuals need to delve into the realm of self to explore it. Education during this period was aimed at the all-around development of the individual's personality. This was achieved through the development of intellectual, physical, moral, and spiritual abilities. Preparation for life and the development of character was another aim of Buddhist education. Students were prepared to earn their livelihood and lead a normal simple life by imparting worldly and practical knowledge. Buddhist form of education believed in cleaning the mind from immoral desires by practicing the eight-fold noble paths or 'AryaAshtangMarg'. These eight-fold noble paths were correct view to develop accurate understanding, righteous thoughts with no intention of attachment, hatred or harming anyone; correct speech that is not harsh, divisive, or meaningless; actions that refrain from physical misdeeds, like killing, stealing, and offensive sexual conduct; upright livelihood that avoids trading that would harm other human beings or animals by either slaughtering, poisoning, and intoxication; doing correct efforts by blocking negativity and harping on positive states of mind; concentrating on presence and awareness of body, mind and thoughts; focusing on eternal pursuit of life.

Nevertheless, during the Buddhist education period, women's education suffered a major setback. Women were not permitted entry to the monasteries; however, later, they were granted permission with many restrictions. Buddhist education flourished to a great extent during the reign of King Ashoka. Thousands of monasteries were set up during this period to impart education. Scholarships, grants, and other such benefits were provided to the students to encourage education.

Takshashila was the most famous Buddhist seat of higher learning, among the other learnings. The universities of Nalanda, Vikramshila, and Vallabhi were great centers for learning and the oldest structured form of universities that existed in ancient India. Nalanda University, situated in Bihar was an institution of higher education and was famous for studies on Buddhism, thus, drawing students' interest to study valuable Buddhist manuscripts from China, Nepal, Tibet, and Korea (Choudhary 2009).

The Buddhist form of education was one of the well-organized education systems that influenced society with the sole aim to bring harmony. This system of education was free from communal impediment, and it focused on the development of society at large.

12.2.3 The Jain Education System

Jainism roughly originated in the 7th–5th century BCE. This was a period of intense religious conjecture and activity. The Jinas or Thirthankars are the founders of Jainism. The Jain belief holds that the Thirthankars are the highest ideal who have shed their ignorance and possess unbounded knowledge, bliss, and power. Vardhaman Mahaveer was the 24thThirthankar and a contemporary of Gautam Buddha who made a major contribution to Jainism. However, according to the Jain belief, he was not the founder of the faith; rather, he was one of the enlightened sages who realized the true nature of reality and the soul. The aim of education according to the Jain system of education was 'Liberation'; the two types of liberation are termed as 'JivanMukti' and 'DravyaMukti'. According to the Jains, education should help the student to perform such karma that will lead them toward liberation. The Jains believed that emancipation from the birth and death cycle could be achieved by practicing the five vows: 'Ahimsa', 'Satya', 'Aparigraha', 'Asteya', and 'Brahmacharya'. An important feature of this system was liberal women's education. During this period, both men and women were allowed admission to the monasteries for education.

12.2.4 The Medieval Period

The medieval period started around the 8th century. This period can be broadly divided into the 'early medieval' period, from 8th century to 13th century, and the 'late medieval' period from the 13th century to 16th century. The Medieval period of Indian history comprised of major social and cultural upheavals and drastic changes. This period witnessed the rise of the 'Mughal Empire' that brought a myriad of changes in society. Some of the major changes that occurred were the decline of the Vedic period and Buddhism established Islam as a dominant force in Indian subcontinent. During this period, the Mahmud of Ghazna invaded India, looted wealth, destroyed the educational institutions, and burned the archives of knowledge repositories. The ancient system of education was replaced by the Islamist education system introduced by the Muslim rulers who established their empire permanently in India.

The foremost aim of education during the medieval period was the propagation of Islam. The basis of education during this period was religion. Muslim education is aimed at gaining material prosperity. In medieval India, there were usually three ways through which knowledge was disseminated. These were Maktab, Madrasah, and Khangah, wherein the elementary education was imparted in, Maktab, whereas Madrasah was a place for higher education, and Khangah was a place for religious education where Islamist indoctrination was discussed. This was also where Sufism got embedded into Islam (Ghosh 2001). During this period, the 'Purdah' system was strictly followed wherein women were not permitted to go in front of outsiders without covering their faces. The 'Purdah' system put restrictions on women's education (unlike the ancient period), women were treated as secondary citizens, and women's education did not gain recognition during the medieval period.

12.2.5 The Colonial Period

The establishment of British colonies brought tremendous changes. Education was one of the major systems that was subjected to radical change. Before the Britishers invaded India indigenous education was prevalent in the country, which was mostly religious and literary in nature. By the 19th century, India was completely under the British rule and the English system of education was encouraged. Consequently, the indigenous education started witnessing a downfall. During the period from 1835 to 1853, the Indian education system saw tremendous upheavals. This period was marked by the 'Orientalist–Anglicist'debate concerning the issues of medium of instruction. There were Orientalists who opined those classical languages like Sanskrit, Persian, and Arabic should be the medium of instruction. In opposition to this view were the Anglicists who wanted to promote western ways of education in India by supporting English as the language of dissemination. Lord Macaulay played a vital role in setting the education system, in part, by putting an end to the Orientalist–Anglicist debate.

The next major milestone in the development of education in this period was the Wood's Dispatch of 1854, as it put forth recommendations for the improvement of the education system in the country. It was Sir Charles Wood who brought about significant impact on Indian education by laying down unequivocal guidelines for education.The Despatch made major recommendations and clearly expounded on the aims and objectives of instituting the grant-in-aid system. It also emphasized the formation of the Department of Public Instruction in the five residencies of Bengal, Bombay, Madras, the Northwestern Province, and the Punjab province. The government was directed to design a plan of education from primary to university levels and finance them, and suggestions were made on improving the indigenous schools. The Despatch made major recommendations for establishing teacher training institutions and bringing the professional studies like law, medicine, and engineering under the control of universities. The importance of women's education was stressed. The Wood's Dispatch was considered as the 'Magna Carta' of Indian education.

12.2.6 The Indian Education Post-Colonial Period

The Indian education system during the colonial period was largely dominated by the British. The initial years after independence were crucial in Indian history, as the country witnessed unforeseen difficulties during this period. Consequently, education remained a neglected sector until 1951. It was only in 1950–51 that the government of India embraced a strategy of planned development that gave rise to the 'First Five-Year Plan'for the development of the country. Thereafter, consecutive plans were executed, wherein the primary educational goals of these plans were to achieve universal elementary education, abolish illiteracy, initiate occupational education programs, and provide facilities for improving the quality of educational system.

The government of India assigned the task of reforming the Indian educational system to various commissions. The first commission was the 'University Education Commission 1948–49'which made significant recommendations regarding the restructuring of the courses, evaluation methods, medium of instruction, student engagement, and the staffing of teachers. The commission proposed that the basic educational aim should be to invigorate the innate abilities of the students, and train them for self-development and democracy. Value inculcation and acquainting the students with cultural heritage was another important recommendation by the commission. Moreover, the commission recommended that admission to the university course should be in accordance with that of the intermediate examination. The commission also emphasized that students should be acquainted with their physical environment, motivated for effective use of language, and the introduction of the basic ideas of science.

The University Education Commission was followed by the 'Secondary Education Commission 1953–1954'. Although there were abundant secondary schools even during the colonial period, the schools were depleted in quality and unable to provide to the changing socioeconomic requirements of the existing societal structure. Hence a strong need was felt to reform secondary education. The University Education Commission also highlighted the fact that our secondary education was frail, and a reformation was needed urgently. Therefore, in January 1948, the Central Advisory Board suggested appointing a commission to examine the existing Secondary Education and proposed measures to reorganize and improve the secondary education system. The major recommendations made by the Secondary Education Commission regarding the educational aims encompassed the promotion of democratic citizenship, upgrading of vocational efficiency, enhancement of students'character, and providing education for leadership. The commission stressed the organization of secondary education by recommending that the duration of secondary education should be 7 years, the lower secondary stage should be 3 years, while the higher secondary stage should comprise of 3 years. Another key recommendation made by the commission was regarding the abolition of the present intermediate classes and establishing multipurpose schools to teach technology, commerce, agriculture, fine art, and home sciences.

Later, the 'Indian Education Commission 1964–1966' widely known as Kothari Education Commission was established by the government and chaired by prominent educationist, Dr. D. S. Kothari, to examine different aspects of the education system prevailing in India. After a meticulous study, the commission submitted its report on Indian education with major recommendations which included equal opportunity for the underprivileged section of society to use education as a tool to improve their social and economic status. Provision of free and compulsory education, adoption of the common school system in India, and provisions for the education of physically and mentally handicapped children were other highlights of the Kothari Commission's recommendations. The commission recommended a uniform educational structure, i.e., 10 + 2 + 3, and the 'Three Language Formula' to be followed throughout the country. The recommendations of the Kothari Commission were accommodated in India's first National Policy on Education in 1986.

In 1986, the government of India, with the aim of improving the educational system, declared its policy on education and titled the document 'National Policy of Education 1986'. Later in 1990, a committee chaired by Acharaya Rammurti was set up to review the National Policy of Education (NPE) to bring educational reforms. Based on these recommendations, Shri N. Janadhana Reddy, Chief Minister of Andhra Pradesh, made a few revisions to the 'National Policy of Education 1986' and submitted a report in 1992, which is recognized as 'National Program of Action (POA) 1992'.

The foremost aim of the National Education Policy (1986) was to bring drastic modernization of education, based on suggestions by the Education Commission, that was crucial for economic and cultural progress, for its national integrity, and for understanding and upholding the socialistic pattern of Indian society. It stressed transforming education to be more relatable to Indian societal needs, with emphasis on providing educational opportunities for all. NEP highlighted the need for sustainable efforts to elevate the quality of education at all stages, with significant emphasis on the promotion of science and technology and nurturing of moral and social values within the learners. NEP proposed that the Indian educational system must be so strengthened that it could produce youth who are virtuous, have a charismatic character, and are dedicated to upholding national integrity (National Education Policy 1968, as cited by Kattackal 1975).

12.3 NATIONAL EDUCATION POLICY-2020 – A LANDMARK IN INDIA'S HISTORY OF EDUCATION

At present, the entire world is going through the 4th Industrial Revolution, which is the digital era, bringing tremendous transformations to industry and other related sectors. Consequently, the industrial skills requirements and societal needs have changed. Thus, to cater to these changes and prepare individuals to face the challenges of Industry 4.0, the education sector also needed a strong reformation. The 'National Policy of Education 2020' (NEP 2020) is a major step by the Indian government to align the educational structure and complete system with the needs of Industry 4.0. The policy was revised after 34 years to meet the changing societal needs of India, and the world at large, and thus bringing education on par with international education standards.

The policy is all-inclusive, holistic, and far-reaching, having the potential to contribute to the progress of the entire nation. It emphasizes a flexible learner-centered system of education that will strive to transform India into a dynamic knowledge society.

According to Gavhane, the most important goal of NEP 2020 is to provide education that offers universal access and affordability, from the pre-primary level to grade 12, and to realize complete enrollment ratio in the education system by 2030. The most notable aspect of NEP 2020 is that it proposes fewer regulations and more autonomy to the educational institutions, emphasizes innovative teaching–learning and evaluation techniques, strengthens teacher training, and promotes more meaningful evaluation processes with a futuristic viewpoint.

The NEP 2020 in the most unequivocal ways has articulated the aims of education to prepare futuristic learners. The major suggestion in the NEP includes that there should be an increase in public investment in education up to 6% of the GDP of India. The NEP 2020 recognizes the importance of early childhood education and recommendsthat the NCERT will be responsible to develop the structure for the children's educationuntilthe age of 8 years. One of the key suggestions was to substitute the 10 + 2 +3 structure of the school system with the 5+3+3+4 structure, wherein the first three years constitute of pre-school or Anganwadi and rest 12 years of schooling and 3 years of schooling. The document also suggests that achieving universal foundational literacy will be the highest priority of the education system by setting up a national foundation for literacy and numeracy.

The NEP 2020 gives due importance to Indian languages and therefore promotes multilingualism. The policy accentuates mother tongue (or local/regional language) to be the medium of instruction up to grade 5 – if possible, up to grade 8 and further. However, the policy suggests that the 3-language formula will be continued considering the Constitutional provisions.

The policy acknowledges the importance of continuous assessment in education and suggests that assessment should be formative instead of summative. It should be based on higher-order thinking, skills, and competency. The policy document recommends establishing a center to assess, reexamine, and analyze the knowledge for holistic development of all the students, as a benchmark body under MHRD. The policy advocates equitable and inclusive education for the socially and economically disadvantaged groups, with special mention to having gender inclusion funds and creating education sections especially for underprivileged and deprived groups.

The NEP 2020 made valuable recommendations on teacher recruitment, it mentioned that the teacher recruitment will be conducted through vigorous and transparent procedures. For this purpose, it recommends establishing professional standards for teachers, responsibility for which would be shouldered by apex teacher education entities of India.

The accreditation and standard-setting of the schools have been given due importance by the policy. It recommended that to maintain minimum proficiency and quality benchmarks, an independent body would be set up to prescribe standards for the schools across all the states. For this purpose, a framework for school quality assessment and accreditation will be developed in consultation with the stakeholders. The recommendations of the policy are very important for vocational education. It may be noted that vocational education shall be incorporated in the curriculum of all secondary schools.

The NEP 2020 has made noteworthy recommendations regarding higher education in the country, wherein one of the important aspects is to boost the ratio of enrollment in higher education, including vocational education.

The contribution of higher education is crucial to the development of youth as well as society and ultimately in evolving India as envisaged by the Constitution. The policy made major recommendations for holistic education at the undergraduate level by suggesting that the goal of higher education of eminence must be progressive, reflective, accomplished, and creative individual personalities to meet the 21st century and Industry 4.0 requirements.

Regarding the structure of higher education, the policy recommends that the degree qualifications shall be for either 3- or 4-year periods, having flexible multiple exits and entry options with corresponding appropriate certifications. The policy also suggested that a digitalstore for academic credits be set up so that academic credits earned from various recognized Higher Education Institutions (HEIs) will make it possible for the learners to avail themselves of degrees from any HEI, as credits would thusly be transferrable.

The major recommendation made by this document involves the transformation of HEIs into large multidisciplinary universities, which will aim to have 3,000 or more students.

With a view to fostering research and an innovation culture, another body will be established to promote and extend research and innovation across the country. The policy also suggested that open and distance education will play a pivotal role in increasing the ratio of overall enrollment to 50%.

The policy further suggested that special support be provided through scholarships to the students from reservation and underprivileged categories. To ensure this countrywide scholarship portal can be extended, it should aid, nurture, and trace the advancement of students in receipt of scholarships. The policy made valuable recommendations regarding having a common single all-embracing egis for the promotion of higher education. Emphasizing the importance of teacher education, the policy recommended a 4-year integrated Bachelor of Education program. A new and all-inclusive common countrywide teacher education curriculum framework will be developed to cater to emerging needs and concerns of teacher education.

As Industry 4.0 demands a digital transformation, education will play a crucial role in this transformation while technology on the other hand will play a significant role in progressing the educational practices and outcomes. For this purpose, the said document recommends creating an independent body, the National Educational Technology Forum(NETF) which will perform the following functions:

a. Offer impartial, evidence-based assistance to government agencies on technology-based interferences.

b. Develop academic capacities in educational technology.

c. Predict well-planned thrust areas in this domain.

d. Enunciate novel directions for innovation and research (NEP 2020).

The policy further suggested that technology-based learning portals will be integrated across school and higher education. The policy directed the ministry of human resource development to shoulder the responsibility of e-learning needs of the school and higher education, through a committed entity for the developing infrastructure and content that supports digital education and capacity building of teachers. The sole purpose of technological interventions will be to improve teaching, learning, and evaluation processes; support teachers in preparing themselves to teach future citizens who would be highly tech savvy; enhance the accessibility of education to every stratum; and streamline in structional planning, management, and administration encompassing every aspect related to educational processes like admissions, attendance, assessments, etc. (NEP 2020). The NEP 2020 recommends developing software that would enhance the educational experience, thus making it available to all stakeholders.

NEP 2020 focuses on embedding disruptive technologies, as it is the need of the hour, wherein industry is completely based. NEP 2020 clearly states that learning must revolve around artificial intelligence, Internet of Things, virtual reality, augmented reality, robotics, machine learning, block chain, data analytics, etc. which will require skills like cognitive flexibility, critical thinking, and problem-solving to be developed among the learners, so for that purpose the education system should be well-equipped to cater to these industry requirements. Thus NEP 2020 proposes entrusting this task (of being vigilant toward emerging trends in technology) to an independent entity that will keep track of disruptive technologies and respond accordingly.

The policy not only emphasizes the importance of digital transformation, disruptive technologies, online learning, but also suggests necessary steps to be taken for the same. The policy recommends that national educational agencies of eminence will conduct pilot studies for understanding the problems and evaluating the benefits of digital transformation. The NEP 2020 further recommends developing a digital repository of content that will include creation of coursework, educational games and simulations, augmented reality, and virtual reality. The digital divide is duly addressed in the policy. Taking into consideration the Indian scenario, where a substantial section of the population will have either limited or no access to digital devices, the policy recommends the use of existing mass media like television and radio to telecast and broadcast the educational programs. Thus, the policy has made remarkable contributions through its recommendations with a view to having a holistic system of education that will prepare generation Z for the 21st century and Industry 4.0 challenges.

12.4 INDUSTRY 4.0

The Industrial Revolution was an alteration in the manufacturing processes that took place during the 18th century in Europe and the United States. Prior to that, most goods were manufactured using human or animal power. The First Industrial Revolution took place roughly in 1765 with the invention of steam power. This period witnessed the automation and mechanization of textile, agriculture, steam ships, and steam-powered locomotives that greatly affected the social, cultural, and economic standing of the world. For the first time in history, production was mechanized, and it proved to be the greatest breakthrough for increasing human productivity. The Second Industrial Revolution emerged in 1870, marking the birth of electrical energy. Other advancements of Industry 2.0 included the combustion engine and assembly lineproduction. The continued experimentations gave rise to Industry 3.0 or the Third Industrial Revolution that began approximately in 1969. This was the age of nuclear energy and electronics. The invention of the programmable logic controller (PLC) in 1960 was a milestone in the automotive industry. The incorporation of electronic hardware into manufacturing generated the need for software systems to facilitate the electronic devices. Subsequently, the process of automation continuously evolved with advancements in electronics and the information technology (IT) industry, resulting in 'Industry 4.0'.

The Fourth Industrial Revolution, widely called 'Industry 4.0', is the ongoing automation of industries using smart technology. The way people connect, communicate and exchange information changed radically with the outbursts of the internet and the telecommunication industry. This brought a paradigm shift in industry, integrating the physical and virtual worlds. Industry 4.0 is characterized by a new level of interconnectivity, big data, artificial intelligence, machine learning, augmented reality, robotics, block chain, etc.

While the Third Industrial Revolution was all about electronics and information communication technology that was useful for automation of production, the Fourth Industrial Revolution is furthering and enhancing the advancement of new technologies that smudges the fine line between the physical environment and digital world.

12.5 THE KEY COMPONENTS OF INDUSTRY 4.0

12.5.1 Big Data

The term 'big data' is used to describe a set of data that is gigantic and growing exponentially. The huge size of data proves to be one of the major problems in data analysis, e.g., the New York Stock Exchange or social media data. The solution to this is what we call 'big data'. Big data analytics makes it possible to identify relevant information from volumes of data which is expanding in an unprecedented way in a variety of formats and get that data analyzed so as to have insights that will enhance decision-making and boost strategic business ventures.

12.5.2 Artificial Intelligence

Artificial intelligence (AI) is the ability displayed by computer programs or machines to think and assimilate. It falls in the computer science domain and involves developing smart machines that can accomplish tasks that would otherwise require human intelligence. Thus, artificial intelligence encompasses the use of computers to do things that would conventionally require human intelligence. This includes creating algorithms to classify, analyze, and predict conclusions from data. Siri, Alexa, and robot-advisors are a few examples of artificial intelligence making human life easier.

The sub-field of artificial intelligence is machine learning, which is based on the notion that systems can analyze data, identify various patterns, and make decisions with little or no human intervention. Most familiar examples of machine learning can be stated as image recognition, e-mail filters, internet search engines, and so on.

12.5.3 Augmented Reality

Augmented reality (AR) has developed and become more persistent among a wide range of applications. It is a superior adaptation of the real physical world achieved using digital visual components, sound, or other sensory stimuli brought in via technology. Augmented reality effectively provides real-time data permitting human beings to integrate the electronic systems and interact with them in an enhanced way (Mitchell1997). In other words, augmented reality is an interactive 3D experience achieved by combining real-world computer technology. Some examples of augmented reality are IKEA's mobile app, Nintendo's Pokémon Go app, L'Oréal makeup app.

12.5.4 Robotics

Robotics is an amalgamation of science, technology, and engineering with a view to producing machines to either substitute or imitate human efforts. Robotics are extensively used in industriesthat involve cyclical tasks, such as the automobile and other manufacturing industries. Moreover, it is used where the manufacturing environment is prone to hazards. Cleaning bots, lawn trimming bots, hospitality bots, and drones are some examples of widely used robots.

12.5.5 Blockchain

Generating blocks in chronological order for the purpose of saving our digital information is termed 'blockchain'. Blockchains can be created as a boundless expansion of the Internet of Things and build a blockchain ecology. Enormous data generated through the Internet of Things can provide a great deal of space in the blockchain (Zhang, 2019).

As the data saved in a blockchain cannot be tampered with and transactions can be authenticated, it is a safe trading system being used in the banking, finance, and insurance sectors. Bitcoin is an example of blockchain technology that is expanding and influencing the world despite the many negative reports in the media.

Although the advent of Industry 4.0 guarantees high productivity, better flexibility, and agility, real-time transactions, improved quality of life, it also brings along adversities like cyber security risks, disruptions of core industries, ethical issues, unemployment, and inequality.

High productivity is an important feature of any industrial revolution. With the advancement in automation, the productivity rate in every industrial revolution multiplies. Automation also makes the upward or downward scaling of production easier, thus ensuring flexibility and agility. Real-time transactions, whether it is customer care service systems or bank ATMs, demand immediate processing. The improved quality of life can be exemplified by mentioning how simple it has become to book a flight, buy a product, or even make payments.

However, the drawbacks of Industry 4.0 cannot be left unobserved. The boom in the internet industry brings with it the risk of data hacking and manipulation or being used for malevolent intent. Disruptions in core industries are being witnessed every day. Traditional taxis are competing with cabs, the shops are competing with Amazon, and so on.

The Cambridge Analytica data scandal in early 2018 cannot be forgotten as an example of ethical issues arising with the advent of Industry 4.0. Digitization and automation in the Fourth Industrial Revolution are envisaged to be fundamental in changing the nature of work, business, education, and society in the coming decades. These changes may lead to the disappearance of jobs and occupations replacing them with new ones. However, in the past few decades, computers have replaced numerous jobs (Bresnahan 1999). Researchers are now shifting their focus of study to the types of jobs that are likely to be displaced by computerization. In a report, Frey and Osborne stated that about 47% of total U.S. employment is at risk of being automated (Frey & Osborne 2013). Moreover, the threat of increased connectivity and cyber security cannot be overlooked. U.S. Department of Homeland Security and the oil and gas industry did emphasize the fact that increased connectivity brings in an increased risk factor (Piggin 2016). Ongoing technological innovation brings an immeasurable range of opportunities and associated challenges, which demand humans be prepared for the same.

India, a nation impacted by the Fourth Industrial Revolution and keen on riding the Industry 4.0 wave, has taken several initiatives in that direction, like embracing Education 4.0.

12.6 EDUCATION 4.0 –PREPARATION FOR INDUSTRY 4.0

The nature of industry and related sectors has changed radically with every industrial revolution. The Fourth Industrial Revolution is not only a technology-driven age but is relatively power-driven with disruptive innovation to constructively impact every other sector, such as health, business, and education (Xuet al. 2018).

However, the key question in the changing scenario is whether India is ready for Industry 4.0. India being the sixth largest manufacturing country in the world plays a pivotal role in adapting to the Industry 4.0 technology. The 4th IR presents a few challenges, like how to use digital transformations and technologies like robotics, virtual tools, automation, artificial intelligence with social inclusion.The Indian government is not only extending support to the industries but is also taking reform measures for adapting to Industry 4.0. Some of the initiatives taken up by the Indian government arethe National Manufacturing Policy to drive manufacturing share to 25% of the GDP, the National Program on Artificial Intelligence for emergent AI applications, and the Mission on Cyber-Physical. Moreover, in 2015 India launched the 'Internet of Things Policy' (IoT) with a view to building a strong IoT system for the overall development of the country.

States and cities like Andhra Pradesh and Bangalore have taken major initiatives to adapt to Industry 4.0. Andhra Pradesh has taken concrete steps to take advantage of IoT potential in the country, while the first smart factory of India where machines speak with each other is being built in Bangalore. A smart factory, equipped with data exchange in manufacturing and the booming Internet of Things will be a revolution in India's preparation for Industry 4.0.

Industry 4.0 has proven to be a game-changer. The revolution has substantially influenced every sector initiating change in the needs of society.

The education sector is no exception to this global trend. Education has always changed its focus with every industrial revolution. During the First Industrial Revolution, education was much focused on standard approaches of learning; the Second Industrial Revolution sawa shift toward bulk production in manufacturing and standardized testing in education; in the Third Industrial Revolution students were seen as customer learning model. The Fourth Industrial Revolution demands that the horizons of education be broadened and has dramatically changed the way every individual relates to one another, lives, works, and gets educated. Thus, it is imperative that the young learners be prepared in such a way that they are able to engage in a world along with smart machines. And for that, the young learners need to be educated differently from how and what they had learned in the past.

Traditional education through information transfer is no longer a workable form of education. Industry 4.0 demands a replacement for the conventional ways of education, and Education 4.0, today's buzzword, is a response to the needs of Industry 4.0 (Delipiter 2019). Education 4.0 is a deliberate and calculated approach to learning that is aligned with the Fourth Industrial Revolution. It emphasizes the transformation of the future of education using advanced technology, with automation, creativity, and innovation being its foundation.

As cited by Aziz Hussin (2017), there are nine trends related to Education 4.0. First, scholars get an occasion to learn at any time, from anywhere with the help of e-learning tools, forever, with tone-paced literacy. Second, scholars will have opportunities for personalized literacy. Third, scholars will have the choice to learn with bias, programs, and ways grounded according to their own preferences. Fourth, students will be exposed to more hands-on project-based learning. Fifth, scholars will have further hands-on literacy by means of field exposure and opportunities like internships, mentoring systems, and cooperative systems. Sixth, scholars will learn to interpret data by applying the generalities learned through their theoretical knowledge and using their logical 'chops' to draw consequences grounded on sense and trends from given sets of data. Seventh, the conventional approaches to assess scholars may become inapplicable or insignificant as scholars will be assessed uniquely. Eighth, every pupil's opinion needs to be taken into consideration while designing and revising the class. Literacy will be scholar-led, therefore compelling the preceptors to take the mantle of being facilitators and guide the scholars through their literacy process(Hussin 2017).

12.7 CHALLENGES TO EDUCATION 4.0

Industry 4.0 has changed the fabric of the education system. To align with the change, one must revisit and redesign the traditional educational paradigms with a futuristic approach. This involves innumerable challenges. The requirement is for an agile and well-resourced education sector. The need for the technology-integrated education system will require a thriving science and technology community and a buoyant information and communication technology (ICT) sector. As many job opportunities will be available to students due to Industry 4.0, Education 4.0 must align itself with the latest technological tools, like machine learning, deep learning, data analytics, etc. As a result, Education 4.0 will have to incorporate changes in the curriculum structure and pedagogical and transactional approaches in the teaching and learning process. Education 4.0 must focus more on digital skills, social–emotional learning, science, technology, engineering and mathematics.

Equipping the learners with employability skills will be another major challenge for the education sector. Education 4.0 will require a drastic change in the education system, from demand-led education to supply-led education; skill-based instead of knowledge-based, and lifelong learning instead of rote-learning. A technologically integrated education system demands the integration of computer-facilitated engagements; this poses a big trial for teachers. Teachers need to be agile learners themselves to adapt to rapid changes in education; they should keep updating themselves with the latest trends in education and keep abreast of industry requirements. Teachers need to teach their students how to learn, how to solve problems, and what not to learn, thus making their students independent learners. But for this to happen extensive teacher training needs to be conducted wherein teachers need to unlearn, relearn, and e-learn. Only if all the teachers are aligned to adopt disruptive teaching–learning processes, like flipped classrooms, and make use of artificial intelligence strategically will Education 4.0 be achieved.

Keeping in view the changing needs, preparing for the various challenges, and aligning with Industry 4.0, India, after more than three decades has come up with an education policy that will certainly change the framework of the Indian education system. The National Education Policy 2020 with its all-inclusive guiding principles right from pre-primary education to higher education, including vocational education, in both rural and urban India will undoubtedly transform India's education system and prove to be a landmark in the history of Indian education.

REFERENCES

Bresnahan, T. F. (1999). Computerisation and Wage Dispersion: An Analytical Reinterpretation. *The Economic Journal*, 109(456), 390–415. Retrieved from https://onlinelibrary.wiley.com/doi/10.1111/1468-0297.00442

Choudhary, S. K. (2009). Higher Education in India: A Socio- Historical Journey fromAncient Period to 2006–07. *Journal of Educational Enquiry*.8. Retrieved from esearchgate.net/publication/237369663_Higher_Education_in_India_a_Socio_Historical_Journey_from_Ancient_Period_to_2006-07

Delipiter, L. (2019). Education and Industrial Revolution 4.0, 10, 48–62. 10.24114/jh.v10i1. Retrieved from https://www.researchgate.net/publication/334837153

Frey, C. B., &Osborne, M. A. (2013). The Future of Employment: How Susceptible Are Jobs to Computerization? Retrieved from University of Oxford, Oxford Martin School website: https://www.oxfordmartin.ox.ac.uk

Gavhane, S.(n.d.).Report on India's New Education Policy 2020A Critical Appraisal. Retrieved from https://mgmu.ac.in/wp-content/uploads/NEP-Indias-New-Education-Policy_2020-final.pdf

Ghosh, S. C. (2001). *The History of Education in Mediaeval India, 1192 AD-1757 A.D.* Originals, Delhi.

Hussin, A. A. (2017). Education 4.0 Made Simple: Ideas for Teaching. *International Journal of Education & Literacy Studies*, 6(3), 92–98. Retrieved from https://www.journals.aiac.org.au/index.php/ijels

India, National Policy on Education, Ministry of Human Resource Development, New Delhi. Retrievedfrom https://www.education.gov.in/sites/upload_files/mhrd/files/NEP_Final_English_0.pdf

India, National Policy on Education, New Delhi, Ministry of Education (1968). Retrieved from https://www.education.gov.in/sites/upload_files/mhrd/files/document-reports/NPE-1968.pdf

Kattackal, J. A. (1975). Educational Development in the Post-colonial Period in India:Problemsand Prospects. Paper presented at the International Colloquium of the Institute for International Cooperation. Seventh, Ottawa, Ontario, April 8–11. Retrieved from https://eric.ed.gov/?id=ED130932

Mitchell, T. M.(1997). *Machine Learning*.McGraw-Hill. Retrieved from in.ufpe.br/~cavmj/Machine%20-%20Learning%20-%20Tom%20Mitchell.pdf

National Education Policy. (2020). https://www.education.gov.in/sites/upload_files/mhrd/files/NEP_Final_English_0.pdf

Piggin, R. (2016). *Risk in the Fourth Industrial Revolution*. The British Computer Society. Retrieved from https://itnow.oxfordjournals.orgdoi:10.1093/itnow/bww073

Report of All India Management Association. (2018). Industry 4.0: India Gearing up for Change. Retrieved from https://resources.aima.in/presentations/AIMA-KPMG-industry-4-0-report.pdfReportonIndia'sReadinessforIndustry4.0-AFocusonAutomotiveSector.https://www.gita.org.in/Attachments/Reports/India%E2%80%99s%20Readiness%20for%20Industry%204.0.pdf

Xu, M., David, J. M., &Kim, S. H. (2018). The Fourth Industrial Revolution: Opportunities and Challenges. *International Journal of Financial Research*.9(2). Retrieved from https://www.sci-edupress.com/journal/index.php/ijfr

Zhang, P. (2019). Analysis on building a green eco-smart city based on block chain technology. In: Abawajy, J., Choo, K. K., Islam, P., Xu, Z., & Atiquzzaman, M. (Eds.), International Conference on Applications and Techniques in Cyber Security and Intelligence ATCI 2018. Advances in Intelligent Systems and Computing (Vol. 842). Springer. https://doi.org/10.1007/978-3-319 -98776-7_60.

CHAPTER 13

National Academic Credit Bank

Need of Times

Neha Vashistha, Aashish Dhiman,
Simon Grima, and Kiran Sood

CONTENTS

DOI: 10.1201/9781003254942-13

13.1 INTRODUCTION

The University Grant Commission (UGC) has introduced the concept of a National Academic Credit Bank (NAC Bank), which is a digital/virtual/online organization that will be established and regulated by the UGC. The main purpose of the NAC Bank would be to facilitate the transition of students through the education system where debts can be accumulated and will be used to change the time requirements for compliance with part of the grade program. The UGC and the government on their side are trying to liberate Indian higher education from a large number of laws and regulators. The NAC Bank would be another big step in this direction. Autonomy is the dynamism of an institution to change the course structure and the curriculum to adapt to the needs of market forces.

Autonomy in higher education in India cannot be limited to the narrow alleys of academic freedom only to academic teachers and administrators. It must reach the bottom of the pyramid, as well as our students. A student's autonomy offers class flexibility, broad weather for developing skills for the market, and opportunities to participate in other activities of their choice. Autonomy is only real if participation is guaranteed for all interested parties, in particular, that of the students.

The Academic Bank of Credit (ABC) is an educational digital platform created to facilitate the mobility of students between or within granting of degrees by Higher Education Institutions (HEIs). Through a well-framed and designed formal credit recognition system, the accumulation of credits, transfer of credits, and reimbursement of creditspromotes-flexible and choice-based learning and teaching. The ABC platform provides students with the opportunity to register for a unique ABC ID, an interactive dashboard to see their accumulation of credits, and options to start a credit transfer mechanism based on choice. Furthermore, the student's module for self-registration will allow careful identification of candidates who want to verify and transfer their credits according to their needs.

13.2 NATIONAL EDUCATION POLICY (NEP)- TRANSFORMING REFORM IN EDUCATION

The National Education Policy 2020 aims to make major transformational reforms in the education sector in India (Figure 13.1).

13.2.1 Conceptual Framework of a NAC Bank

The NAC Bank is a service provider which is available for the student. It will help and facilitate the integration of different campuses and learning systems, creating the mobility of students within the same university/institution and between institutions of their choice. NAC Bank can help to fully integrate the skills and experiences into a formal credit system which provides a mechanism of credit recognition.

The NAC Bank is a digital/virtual/online entity that will be established and administered by the MHRD and the UGC. NAC Bank can operate as a commercial bank in which the students will be the account holders or customers to which the Bank will provide different services which include authentication and verification of credit. NAC Bank will be linked to the National Academic Repository (NAR). NAC Bank can provide deposit accounts to all students who are studying in any recognized institution of higher education. The academic credits which are received and earned by a student in the course and

FIGURE 13.1 Evolution of NAC Bank 2021.

program will be automatically accredited to the account of the student. After the accumulation and collection of credits at certain levels, a student can collect and redeem credits for any course at any point in time. Here, the academic program is an educational program that leads to the degree, diploma, or certificate.

The NAC Bank initiative will help the students to plan their learning. The National Academic Credit Bank will serve as the online student data store fortheir acquired receivables. This will assist and help students who want to compile their academic documents for work or educational purposes on a single platform. NAC Bank can also be helpful for employers/institutions to recognize and identify the credits which are already gained by enrolled students and all institutional members of the system. All the credits required for certification will be made available through the educational transcription of NAC Bank. Transcripts may also mention specific grades to indicate the level of a student's performance in each module/unit in which they get themselves registered and can also judge the difficulty level of the module.

Furthermore, it is perceived that it can also help students with reduced linguistic barriers, checking the pace of a certain extent and the ability to choose the best places or teachers for their studies.

13.2.2 Significance of NAC Bank

A NAC Bank is a medium via which one can choose one's academic road to earn a degree/diploma/post-diploma. It functions on the basic principle of the multiple entries and multiple exit (MEME) concept, and also about any time learning, anyplace learning, and any level learning. It also provides a plethora of choices for course corrections, if necessary, during learning and period of study. It is a farcry from traditional systems in which one enters a pipe of an educational system at a specific entry point and exits at a fixed finish point. The model will be analogous to a leaky or open-ended pipeline with various openings if the NAC Bank idea is used.

The UGC has launched this endeavor because it has the potential to be extremely useful to the student community. In reality, it will percolate autonomy down to the students. Finally, it will satisfy the student's need for knowledge, give them the freedom to choose and alter academic paths, connect diverse fields, and assist them in acquiring the proper foundations and building blocks for their aspirations.

NAC Bank system, as an open educational system, will play a critical role in broadening access to education and providing more flexible educational possibilities. Many people, including members of the military, do not have formal education beyond high school. Through this system, members of less privileged social groups have received credits, diplomas, certifications, and degrees. However, debates regarding the NAC Bank system reveal an overemphasis on analyzing the system at the micro-level. That is, current talks concerning the NAC Bank system are primarily concerned with determining how and to what degree students are or are not content with the system, regardless of its sociostructural influence.

13.2.3 The Functionality of NAC Bank

NAC Bank will keep track of a student's academic credits it will not take credit course documents directly from students for any course they may be undertaking, but only through higher education institutes, who must make deposits into the students' accounts (Table 13.1).

13.3 CONCEPTS OF CREDITS AND CREDIT TRANSFERS

Credit will be assigned to the students who have shown that they have completed a module or a unit for qualification. To achieve this goal, they must comply with the specific set of learning outcomes for the module/unit or evaluation. The results of learning would describe what a successful student will know, will understand, and could do. It is the achievement of learning that is important for the award of credit, not where learning was carried out or was evaluated or how much time was taken. To be rewarded, the student must satisfy the minimum standard (threshold) or pass the evaluations. The minimum standard envisaged for qualification would have been established in the evaluation regulations of the HEIs. If the student exceeds the minimum standard, they will not get awarded with extra NAC Bank credits since the credit value is set before evaluation. Academic credits can be freely transferable within a given system due to NAC Bank. As students progress through a study program, they can accumulate the credit value of the modules or units that they have completed successfully. Institutions can also make agreements that can allow students to transfer credits they received. Students can do so as part of a study program to another, offered by the same institution, and/or credit transfer when moving from one institution to another. How students can transfer credits would be determined by the institutional assessment legislation.

TABLE 13.1 Functionality for Schools and Colleges

For Schools	For Colleges
From 10+2 to 5+3+3+4: Current 10+2 structure in which policy covered schooling from classes 1–10 (ages 6–16) and then classes 11–12 (ages 16–8) gives way to 5 years of foundational education, 3 years of preparatory, 3 years of middle, and4 years of secondary schooling.	**SAT-like college test:** National Testing Agency to conduct common college entrance exam twice a year.
Multi- stream: Flexibility to choose subjects across streams; all subjects to be offered at two levels of proficiency.	**4-Year bachelor:** 4-year multidisciplinary bachelor's program to be preferred; midterm dropouts to be given credit with the option to complete a degree after a break.
Diluted board: Board exams to test only core competencies; could become modular (objective andsubjective) and will be offered twice a year.	**No affiliation:** Over the next 15 years colleges will be given graded autonomy to give degrees, affiliation with universities to end, so would be deemed university status.
Multilingual: 3-language policy to continue with a preference for local language medium of instruction untilclass 8.	**Fee cap:** Proposal to cap fee charged by private institutions of higher learning.
Bag-less days: Schoolstudents to have 10 bag-less days in a year during which they are exposed to a vocation of choice.	**Going global:** Top-rated global universities to be facilitated to come to India, top Indian institutions to be encouraged to go global.

To illustrate the characteristics of the NAC Bank, two cases have been considered which are as follows.

Case 1: A student wanted to study physics as his specialization(major). However, with the pressure of parents and peer groups, he was forced into engineering. Although he was a good scorer, he was not happy with his branch. Fortunately, in his first six months, he got an opportunity ata good university in Canada througha studyabroad program. He has the opportunity to work with a very distinct teacher who was a student of a Nobel laureate. On the advice of his mentor, he ended up enrolled in some unique courses such as

(a) The Science of the 20th Century

(b) Quantum Computing

(c) The World of Music. These were in addition to the project activity he performed. The student returned to India with great enthusiasm for and clarity of his path of study and his future. This has been achieved due to academic flexibility and credit transfer structures and facilities.

Case 2: Another student entered a similar engineering discipline to the previous case, but with interest. However, during his studies, he became dissatisfied and triggered with engineering as a career option, and he decided to migrate to the BBA program.

But there were no options and, consequently, he lost a whole year and the conventional system did not help provide any credit transfer, etc. A scheme like NAC Bank could easily help such students in transferring credits to their new degree course, in the form of language course, environmental, or computer introduction course, or others could have converted in the form of open options.

In thiscase, the loss of time could have been reduced to a minimum, and the transition-could have been facilitated for a needy student with the help of NAC Bank.

13.3.1 Objectives of the NAC Bank Scheme

The universal curriculum or graduate specializations for the whole country, which are designed by central authorities or bodies, can make for a limited and monotonous education, which would notbe useful for global competition. The following are the main objectives of the NAC Bank scheme:

- Promoting the centrality of students in higher education.
- Allowing and empowering students to select the best courses/combination of content courses.
- Students can select the best departments or institutions, or their combinations, to meet their interests.
- It makes it feasible and possible for the students to pick a pattern for their studies with attached or incurred costs, etc. Hence, it allows the students to pick their titles/courses or alter specific changes/specializations instead of going to regular prescribed titles/courses.
- Allow multiple entries and multiple existing systems so that students complete their titles and degrees based on their time preferences. Allows mobility between Title/Diploma/post graduate (PG)diploma/PG programs.
- Allow the application of teaching–learning activities with greater mobility or integration of campuses.
- Facilitates learning of life among all formal and informal students, whether in full-time or part-time mode.
- Allows students to learn different aspects of work, practice, culture, environment, etc., during their studies.

NAC Bank will not provide any awards, degrees, diplomas, or certificates. The bank will perform three functions namely,

credit accumulation,

credit transfers, and

credit redemption.

Opening an account with NAC Bank is not mandatory, rather it will bebased on the student's decision.

13.3.2 Stakeholders of the NAC Bank Scheme

- Students – both national and international, parents, institutions of higher education scheme in India (HEIs).
- Industry.
- Faculty members/teachers.
- Statutory bodies and governments, like MHRD, UGC.
- Non-government organizations.

13.3.3 Key Elements of the NAC Bank Program

To take full advantage of these features, universities/institutions should also establish a program to include credits earned through shillings, formal education, and other learning elements acceptable to them. National programs such as NPTEL, SWAYAM, and SWAYAM PRABHA can and should also be taken into consideration for credit. The concept of NAC Bank is by no means limited bydistance or non-contact mode. It may be a combination of various existing modes;the intent is to support teaching–learning. It has the potential to eliminate the stigma associated with distance learning and workplace learning. New generation of students is more satisfied with flexibility and portability. With the emergence of basic technologies that have the potential to provide expected functions and update educational methods, it is time to consider implementing these technologies in our country. This will eventually open the door to infinite possibilities for capable educational institutions and benefit those seeking higher education.

Functions of NAC Bank- The main proposed functionsare:

- Credit accumulation.
- Transfer of credit.
- Credit Redemption.
- Opening, closing, and verifying accounts.

However, NAC Bank does not award any degree or diploma, etc. Universities and autonomous institutions have the power defined from time to time it should only contain credible credit records. In addition, the students don't need to open or select NAC Bank schemes or modules (Figure 13.2).

13.4 WORKFLOW OF NAC BANK

13.4.1 Registration

- In step one, students need to register with their mobile number through DigiLocker (www.abc.digilocker.gov.in) (Figure 13.3).

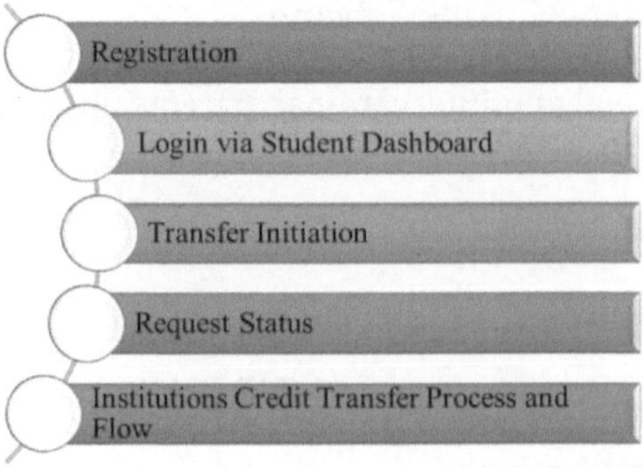

FIGURE 13.2 Workflow of the NAC Bank.

FIGURE 13.3 Student Dashboard. (Source:https://www.digilocker.gov.in/dashboard)

- Log in as a student and click on 'Sign up with DigiLocker'. Student is required to enter a valid mobile number after which a one-time password (OTP) will be sent via SMS. The student enters the OTP and clicks on the 'Continue' button.
- In the next step, the student is required to enter the security PIN (created during sign-up) and click on the 'Submit' button. And, then the student will be directed to the ABC student account creation window.
- After the registration, a unique ABC ID will be allotted to the student.
- In the next step, ID can be shared with all the academic institutions inwhich the student is enrolled.

13.4.2 Loginvia Student Dashboard

- Student logs in with an OTP from a registered mobile number. The student clicks on the 'Go to Dashboard' button – ABC will try to fetch any ABC-linked credit data, but in the case of a new user, it shows a message 'Nocredits added yet' window (Figure 13.4).
- Next, the student provides this ABC ID to the corresponding academic institution where they are enrolled in order to pursue their respective course. It also enables the academic institution to upload credits against student's ABC ID on Digilocker – National Academic Depository.
- When the academic record, along with the credit score of the student, is uploaded by the academic institutions, the credit score will be reflected in the ABC account of the student. Credits mapped to ABC ID will be displayed on the dashboard.
- In the 'Credit Accumulation' window, detailed subject-wise credit distribution is then visible.

13.4.3 Transfer Initiation

- From their dashboard student can select the credits of his/her choice (Figure 13.5).
- On the home screen, a 'Transfer Credits' button is available. The student initiates transfer by selecting the beneficiary instant.
- Students need to click the checkbox to agree with the terms and conditions in order to proceed with the transfer.

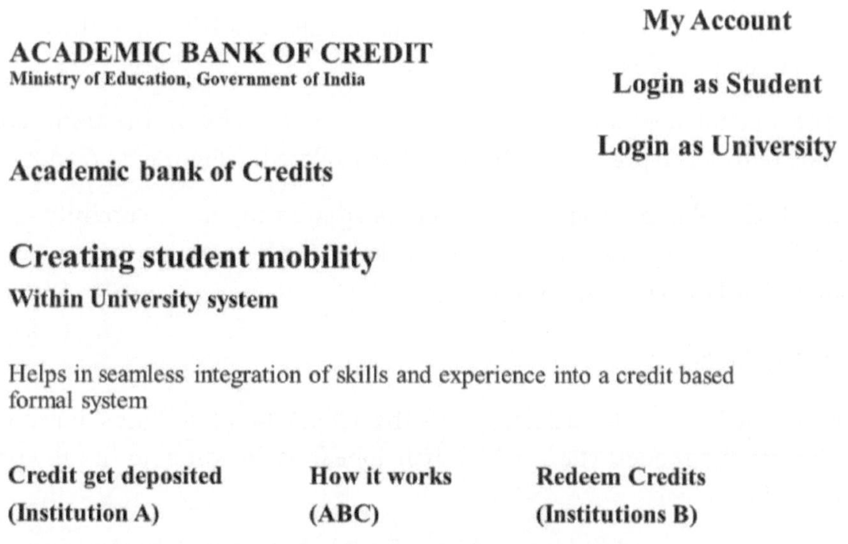

FIGURE 13.4 Student Dashboard. (Source: https://www.abc.gov.in/)

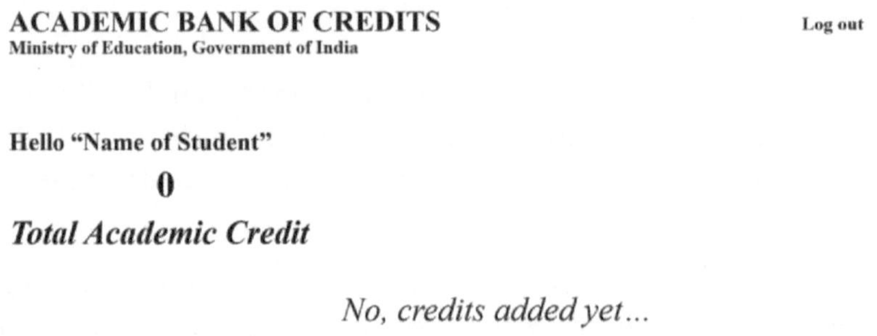

FIGURE 13.5 Login Dashboard. (Source: https://www.abc.gov.in/)

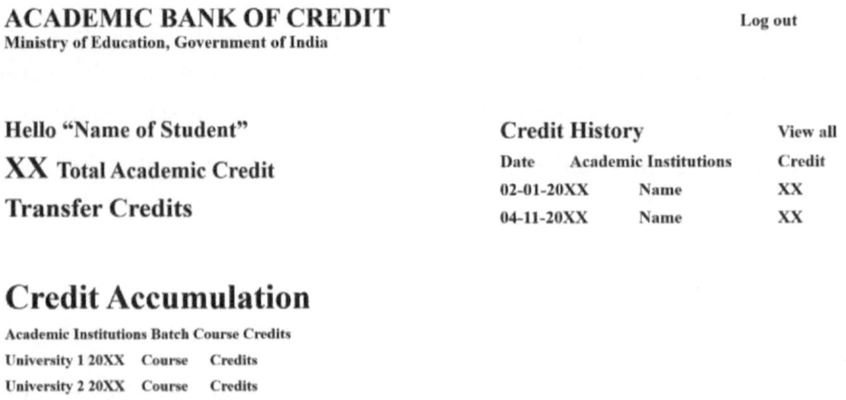

FIGURE 13.6 Total Credits. (Source: https://www.abc.gov.in/)

- Provides other necessary details for transfer like 'University to transfer your credit', 'Course details' and 'Stream detail'. Then the student can proceed to credit selection.
- The student selects the credits of the subject he/she wishes to transfer by clicking on the checkboxes. Here, multiple selections are also available, and then the student proceeds to credit transfer.
- Next the student reviews the subjects selected and clicks on the terms/conditions checkbox in order to proceed with the transfer of credit (Figures 13.6 and 13.7).
- A dialog box will appear, and the student is advised to read it carefully before proceeding further. Once done, the student clicks on the Request tab and the request will be sent with a Transfer Request ID.

13.4.4 Request Status

- On the Home Screen, the student clicks the 'Check Request Status' button and can view the transfer request status which will appear as 'Pending' until fully processed. (Figure 13.8).
- By pressing the expand button on 'Transfer Request Status', students can view the details (Figures 13.9 and 13.10).

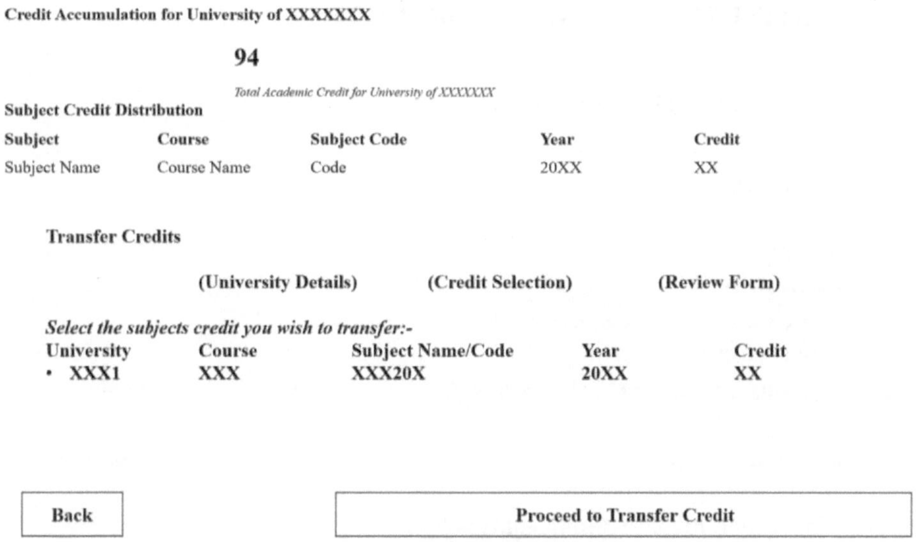

FIGURE 13.7 Credit Transfer. (Source: https://www.abc.gov.in/)

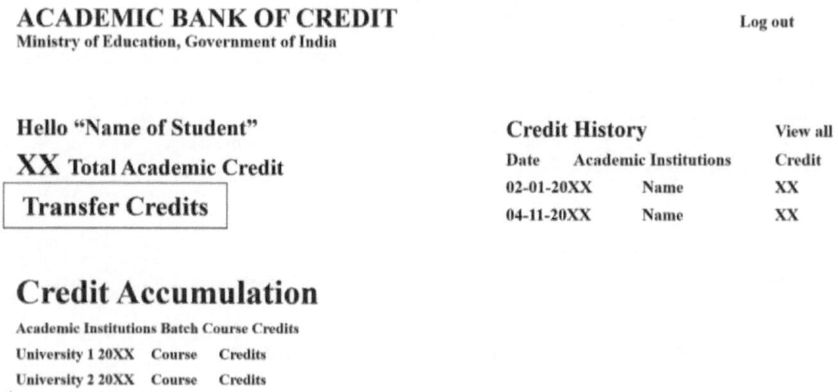

FIGURE 13.8 Student Credit Score. (Source: https://www.abc.gov.in/)

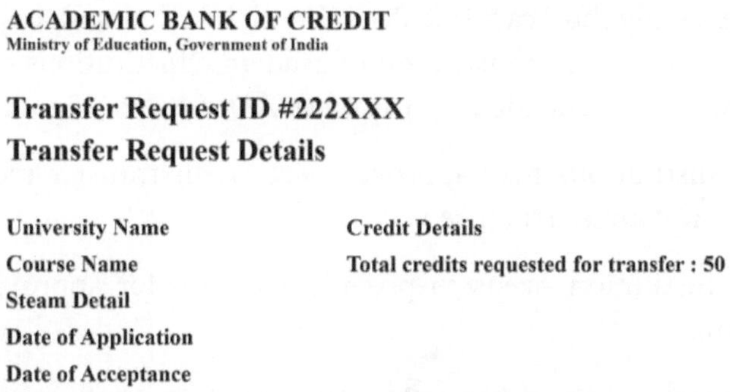

FIGURE 13.9 Transfer Request Approval. (Source: https://www.abc.gov.in/)

INSTITUTIONS NAME, ADDRESS

TRANSFER ID: 12XXXX4

Student Details

Name: Date of Birth:

ABC ID: Gender:

Course Details

Course Name: Stream Details

Credit Details

University	Subject	Subject Code	Year	Credit

Reject Transfer Request **Accept Transfer Request**

FIGURE 13.10 Institution Request. (Source: https://www.abc.gov.in/)

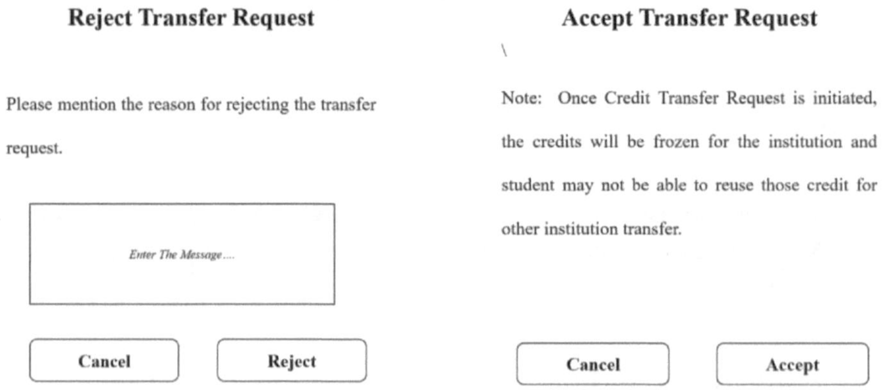

Reject Transfer Request

Please mention the reason for rejecting the transfer request.

Enter The Message....

Cancel Reject

Accept Transfer Request

\

Note: Once Credit Transfer Request is initiated, the credits will be frozen for the institution and student may not be able to reuse those credit for other institution transfer.

Cancel Accept

FIGURE 13.11 Transfer Request. (Source: https://www.abc.gov.in/)

13.4.5 Institution Credit Transfer Process and Flow

- Academic institutions (AI) check the eligibility of the transfer request based on the UGC guidelines. Only the Academic Bank Credit institutions can process the request to transfer the credits. The registration of academic institutions can be done on the DigiLocker National Academic Depository (NAD) system (Figure 13.11).
- The academic institutions may approve/reject credit transfer requests partially or fully based on the course structure.
- The academic institution needs to provide a reason for approval/rejection in the remarks column.
- All the requests related to credit transfer and the corresponding status will be visible on the dashboard (Figure 13.12).

INSTITUTIONS NAME, ADDRESS

ABC ID: 0102XXXXXX

Students Details
Name:
Date of Birth:
Gender:

Credit Accumulation

Academic Institution	Batch	Course	Credit

FIGURE 13.12 Student Credit Information. (Source: https://www.abc.gov.in/)

13.5 IMPLEMENTATION METHOD

NAC Bank operates under a standard framework supported by UGC, under a standard framework like NAD (National Academic Depository). The scheme provides for the transfer of credits from one institution to another, from one department to another, and from one major or specialization to another, to facilitate the transfer of students between institutions and majors from time to time. This will allow them to build their degree the way they want, but will be within the scope of the standard course structure, as well as the flexibility provided as part of the overall course. A system must essentially be a credit-based system to allow the interoperability and transfer of credits from one institution to another. Participating institutions/universities must formally obtain qualifications and agree to participate in the NAC Bank scheme. In a case like this, institutions must change their regulations and policies in terms of course registration, course admissions, course requirements, fees, attendance, credits granted, type of grades, and admission to interdisciplinary courses. It should also be formally approved by its internal institutions, such as the Board of Studies and the Academic Council/Senate. If necessary, the approval of other regulatory agencies such as MCI, BCI, PCI, NCI, etc. should also be given priority.

The credits earned are stored/recorded in recognized national institutions and have a fully defined shelf life for the value of awarding degrees/diplomas/PG diplomas/certificates. Once credits are used for rewards, they will be deducted from individual student files/records. Based on the total credits being earned and the type of credits earned, students can make themselves eligible for a degree/diploma/PG diploma/certificate, authorized according to the regulations of the host/parent institutions from where the student seeks the award. For this reason, it is recommended that students receive at least enough percentageof credits from their parent institutions and would be eligible for awards. It is recommended to do so because it allows students to understand the culture of diverse institutions, and be able to carry out group activities and teamwork. He/she will also have to only undertake project activities at the parent institutions from where the final degree will be obtained.

The development or completion of academic performance is carried out using variable credits and transfers between departments and inter-departmental or internal and inter-departmental institutions. Credits must be obtained from a recognized accredited/acceptable institution and be part of such a program. As the project/thesis work of different institutions may have very different credits/weights, it is recommended that only the parent institutions can meet such requirements (if any). The proposed NAC Bank scheme should be based on the basic principle of promoting more interdisciplinary courses and to grant more credits for elective courses and open elective courses. The learning structure (hence the credits) can be divided into mandatory subjects/core subjects/optional courses/open elective courses, etc., or it can be divided into other different categories according to the needs/decisions of different majors. The plan to activate NAC Bank should be approved by the relevant Senate/Academic Committee or other such authorities of the university.

13.6 ACADEMIC BANK OF CREDIT: A SILVER BULLET FOR THE INDIAN HIGHER EDUCATION SYSTEM

The initiative of Academic Bank of Credit has the potential to globalize and internationalize the Indian Higher Education system. In many other countries, like Canada, the UK, South Korea, such academic credit banks are already functioning. The University Grant Commission sampled this idea first in 2019, then adopted it in 2020 as New Education Policy after discussions and elaborations. In India, the Academic Bank of Credit is a gazetted notification and regulation. The higher education institutes have agreed to implement and join it. The University Grant Commission has communicated it to the vice-chancellors of higher education institutes. This facility has provided the virtual space to the students by which they can deposit and assemble earned credits during their courses that areoffered by the host or Indian universities. This initiative of Academic Bank of Credit is student-centric which offers a friendly learning approach andensures a multidisciplinary holistic education. This initiative will enable the students to pick and select the best course or combination of courses that suit the aptitude and quest for the knowledge of the student. By availing such choices, students feel motivated and read the papers which attract them. The ABC allows the students to alter and tailor their degrees and make the specifications and modifications in the courses, they are no more rigid thanread-only courses being offered by the universities. It provides multiple entries and exits for the students;this freedom of mobility also will assist in reducing the dropout numbers and increase the enrollment graph of the students in higher education.

In India, the education system is diversified, and in-country there are central universities, State Universities, and deemed universities which have full, partial, or no support fromthe University Grant Commission. The degree-granting mechanism and credit allotment and transfer system vary from state to state, from university to university. Before the ABC scheme, when the student moved from one place to another in the country, they faced problems if they were in the middle of their program. The ABC will reduce such problems for the students as well as for the institutions. This will also enhance the pedagogical level of the higher education institutions of India. The European Credit Transfer system recommended that the European Higher Education become more transparent and assiststudents to move within the country and go abroad for education anytime during the session. Similarly, the South Korean Academic Credit Bank System (ACBS) states that it makes it possible for the students to avail themselves of and earn a degree by combining different courses of their choice. It is advantageous for the students, as they can now without any fear or difficulty complete their degrees. They are also relieved of any repeat classes. The same status is also being presented by Canadian and American institutions.

In the global scenario, nations are moving forward in thesharing of knowledge and innovation. As a developing nation, India cannot work in isolation by only following a traditional and rigid system of allotting and granting degrees. The ABC regulation will standardize the Indian Higher Education System to global and international levels.

13.7 CONCLUSION

The proposed initiative has several strong merits to open and strengthen the Indian higher education system during the strengthening of all stakeholders to provide a more open education in the center. Due to the support for multiple entries and multiple exits, it has the potential to increase the 'gross enrollment ratio'beyond the great expectation of 5% in the next decadeand be a key enabler for lifelong learning. It will enable the students to choose the best combination of courses as per their interests, asthey will now not be bound to study those courses and subjects only in which they take admission traditionally. Students can select the best of the departments or institutions or a combination of both. This system will give pace to the studies of the students;also, they will be able tomodify their specializations and their degree program rather than going for a regular degree program or courses. The NAC Bank also provides mobility across various disciplines whether in degree, diploma, PG diploma, or certificate program. This will also maintain the academic record of the students:they will also get the transcript issued, and it will be recognized by all the memberuniversities. The NEP 2020, which replaced the National Policy on Education 1986, is a framework that focuses on the elementary level of education to higher education in India. This scheme will universalize education from pre-school to the secondary level.

The UGC may have to undertake the following:

- Create a special group/body to facilitate the implementation of NAC Bank throughout the country.
- Carry out a survey or meeting with the stakeholders before it is launched, to validate the ideas and seek critical voices and inputs.
- Perform a series of work/training programs in all regions of the country to educate the main academic officials on this scheme and its implementation.
- Review the scheme completely after two years to make structural changes if any.

BIBLIOGRAPHY

European Credit Transfer & Accumulation System, https://ec.europa.eu/education/resources-and-tools/european-credit-transfer-and-accumulation-system-ects_en

https://www.google.com/amp/s/indianexpress.com/article/explained/reading-neweducationpolicy-india-schools-colleges-6531603/lite/

https://www.nad.gov.in/

UGC: Approach Note by Dr. Bhushan Patwardhan, Vice-Chairman, UGC, New Delhi January zOlg.

US & Canada Academic Credit Stacking System — Personal Discussions with some experts.

Usher A., *The Korean Academic Credit Bank: A Model for Credit Transfer in North America, Higher Education Strategy Associates Intelligence Brief 8*, August 2014.

CHAPTER 14

Technology in Education vs Teacher-Taught Relationship in the 21st Century

Ram Krishan

CONTENTS

14.1 INTRODUCTION

Traditional formal education has witnessed a long journey from many generations past. Although learning is a lifelong process, formal education has always excelled at the pace of progress of society. The pious relationship of teacher and taught has paved the way for society to be more civilized and progressive. The teacher–taught roles, relations, and the prevailing value system along the process of teaching and learning are greatly affected by the commencement of technology. The reorganization of all education parameters due to technology is going to influence everything (1). For instance, socioeconomic status and the resultant quality difference between private and public institutes has created an uneven results in different parts of India (2). Primarily technology entered education as part of the curricula, an interactive tool for classroom teaching to enhance learning (3). The focus is learning providing freedom to students, with the teacher taking the back seat to facilitate the constructive environment for inculcating problem-solving abilities in them. In 21st century, the impact of technology has resulted in easy and large availability of information, interactive

DOI: 10.1201/9781003254942-14

teaching, easy sharing of knowledge, and an increase in interest of the learner. The right blend of technology with education can do wonders, as the learning environment becomes more stimulating, and the process of education can be made more interesting. Subramaniam (2016) concluded that although the process of teaching could be altered by the involvement of technology, even then the teachers' knowledge should lead the techno practices (4). The technology has brought a huge flow of information which along with imparting knowledge also creates turbulence due to information pollution and new moral challenges. So a teacher's role is very significant to strengthen the students morally to prioritize wisely. Thomas Lickona (2004) has emphasized thinking about character and academics simultaneously (5). Moreover, in the present scenario, it has become very important for a teacher to groom the students at life and professional skills to survive in the competing world.

Information technology generally focuses on software-based solutions. There is no strong mathematical or computational background required. Artificial intelligence aims to develop intelligent systems that think like humans; and teaching AI should include exposure to philosophy and psychology of humans and different fields of science, in particular computer science.

14.2 TECHNOLOGY IN EDUCATION

The information technology (IT) sector in India is leaping forward to the latest technologies viz, nanotechnology, artificial intelligence, IoT (Internet of things), so academia has to implement requisite modifications in teaching–learning to update and upgrade students. Technology in education has initially been used in distance education, as a tool in classroom teaching and online learning management systems.

- The National Education Department of Higher Education (MHRD, Government of India) National Mission on Education through information and communication technology (NME-ICT) has provided a platform for the people of India from different backgrounds and cultures to explore and exchange information and ideas for two decades. It aims to broaden the periphery of education to anyone in any place by generating online learning portals and virtual tools (6).
- The emphasis of the National Education Policy (2019) focuses on the rigorous technical professional growth of teachers, ensuring access to education to all, irrespective of any disparity (7).
- The potential of artificial intelligence gets highlighted in the discussion paper of NITI Aayog (June 2018) as #AIforAll as a brand for India, proposing transformations in the health, education, agriculture, and urban development sectors (8).
- In education, intelligent tutoring systems that are AI-based, computer-supported teaching–learning environments are the basic model (9). The applications involve sharing content, seeking feedback from students, setting exams, grading, and many more, which are progressing in developed countries and is still in infancy in India (Table 14.1).

TABLE 14.1 Traditional vs Online Education

Traditional Education	Online Education
Advantages	**Advantages**
1. Face to face contact between teachers, students, and advisors	1. More flexible for students who work around the clock
2. More connection to campus activities	2. The capability to complete assignments at times when it is most comfortable
3. More convenient for hands-on training	3. The capability to obtain a certificate from a school in another state
4. This model of education is perfect for all fields	4. This method of education is perfect for army students
5. More travel	5. Less travel
Disadvantages	**Disadvantages**
1. Less flexibility in the class schedule	1. Low direct touch with students and teacher
2. Less cost-effective	2. Limited time for student and campus activities
3. More travel to and from the institute	3. Subject to 'technical complications' and software crashes

14.3 ONLINE LEARNING MODULES

At the international level, Open Software Systems have been very popular for a long time, and although the Indian education system also adopted this software according to the suitability of the education aspirants, adoption is moving at a very slow pace. Nevertheless, growth in the number of students has been observed especially since the sharp rise in internet users mainly on smartphones and other handheld devices. We can take the case of NPTEL, SWAYAM, and IIT, pioneers in spoken tutorials delivering open courses suitable for Indian students (10–12).

14.3.1 NPTEL

NPTEL is an abbreviation for National Program on Technology Enhanced Learning which is run by seven Indian Institutes of Technology (IITs) for making course material in design and science. NPTEL began during the years 1999–2003 after several discussions between IITs, IIMs (Indian Institutes of Management), and CMU (Carnegie Mellon University). A proposal was mutually signed by five IITs and IISc to make electronic content and 100 forty-hour video courses.

14.3.2 SWAYAM

The Study Web of Active Learning by Young and Aspiring Minds (SWAYAM) program is run by the Indian government with the aim to achieve the three important standards for education:

1. Access to education
2. Value of education
3. Quality education

The main aim of this program is to provide the best teaching–learning skills to all. Courses communicated through SWAYAM are simple and available at no cost to the students. SWAYAM courses are designed and run by the Ministry of Human Resource Development

(MHRD) and All India Council for Technical Education (AICTE) with the assistance of Microsoft. This portal facilitates around 2000 courses for the students of school, undergraduate, post-graduate, and also provides various expert courses in diverse fields.

14.3.3 IIT

Indian Institutes of Technology (IITs) spoken tutorials started its operations in 2011. The total number of training/workshops conducted since 2011 is 68,729 for 3,514,891 participants in 3770 institutions from all over India. The learning model of IIT spoken tutorials is unique in the way that the students enroll at the institution, get admission in the regular course, and complete the tutorials under the supervision of a local trainer – and get online help too. This is an advantage, as they do not need to spare extra time for practice at home/hostel after regular classes. Another advantage is that the courses offered sync with the regular courses they are supposed to clear in a particular year/semester.

The following Table 14.2 provides the detailed comparison of different online learning modules on the basis of various features used by these portals.

14.4 TEACHER–STUDENT RELATIONSHIP

The relationship among educators and students is the most essential relationship during the time spent training; furthermore it is a unique sort of social relationship, which is vital to student's academic, social, and motivational progress, altogether influencing the educational climate (13).

Figure 14.1 depicts the teacher–student relationship model and different aspects. Great teacher–student relationships, described by undeniable degrees of closeness, assumes a functioning part in the improvement of the two – teacher and student – which adds to students' social-passionate conduct and scholarly change (14, 15). These studies give understanding into how teacher–student connections impact academic results (16, 17). Obviously, the teacher–student relationship lies at the core of teaching and learning; however, it is not difficult to explain and quantify the relationship better according to the viewpoint of teacher and student or both (18). The collaboration among teachers and students is a mind-boggling framework which will include age, sex, family financial status, sentiments, convictions, and different components. At different stages, starting with fundamental education to higher education, the teacher–student relationship will have various qualities, the manners in which how to construct a decent relationship among teachers and students will be unique. Indeed, students' qualities, including scholarly accomplishment, add to not just the arrangement of teachers' inclination and acknowledgment of students but also the academic results of the teacher (19). This is particularly obvious in universities and colleges. The performance in their academics is vital to students, logical proficiency, or examination capacity is more significant. These students join some professor's research activities to make some exploration with the researchers, which will likewise work on the teachers' scholarly performance. Numerous hypotheses and points of view can be applied to concentrate on the teacher–student relationship, theory of attachment (20, 21), relational theory (22, 23), and phenomenological viewpoint (24).

TABLE 14.2 Comparison of Online Learning Modules

Portal Features	Spoken Tutorials	NPTEL	Swayam
About	Spoken tutorials based on FOSS tools	Videos and web content are available in all engineering and science disciplines	Lecture series videos specifically arranged learning material which is available to print or download
Target Group	All who are interested in grasping new technology	All who are interested in the science and engineering qualification at graduate and post-graduate levels	All who are interested in streams of engineering, science, management, mathematics, humanities, language, arts, and commerce, education, etc.
Content Available	The videos are available in more than 22 languages	English records of videos of NPTEL courses are being completed in ten local languages	Video, animation, PowerPoint presentation available in eight Indian languages
Easy to use	Content is available on the website	Content and videos are available on the YouTube channel	Content is available on the website
Technology used	All content is available only on the FOSS tools. No content is available on any proprietary software	All content is approved by AICTE and prepared according to UGC guidelines on e-content writing	Massive open online course (MOOC) platform
Institutes involved	Above 5000	Above 3700	203 partnering institutes
Cost	Free	No cost except the online test registration fee	Free of cost
Utility	Beneficial for the new users	Mostly used by the science and engineering students to gather extra knowledge	9th standard to post-graduate level students get benefits of digital learning from this portal
Certifications	Certificate provided after course completion	Registration is compulsory if a learner wants to get certification after completion of the course. The certification exam fee is Rs. 1000.	Eligibility for the certificate is on the course page, and learners will get certificates if criteria are matched

14.5 CHALLENGES AND POSSIBILITIES

These technological advancements in the teaching–learning environment pose new potential and challenges for teacher–taught relations and the education system, as education is pivotal to humanism – and not to machines. Studies have shown that the proficiency of teachers plays a key role in the achievement of even student-dominated strategies (13). The qualities, like imagination, innovation, creativity, motivation, aesthetic sense, wisdom to choose the right priority, are not machine-made but rather involve human touch. So, the accomplishment to enhance education with technology can become feasible if the challenges are well addressed.

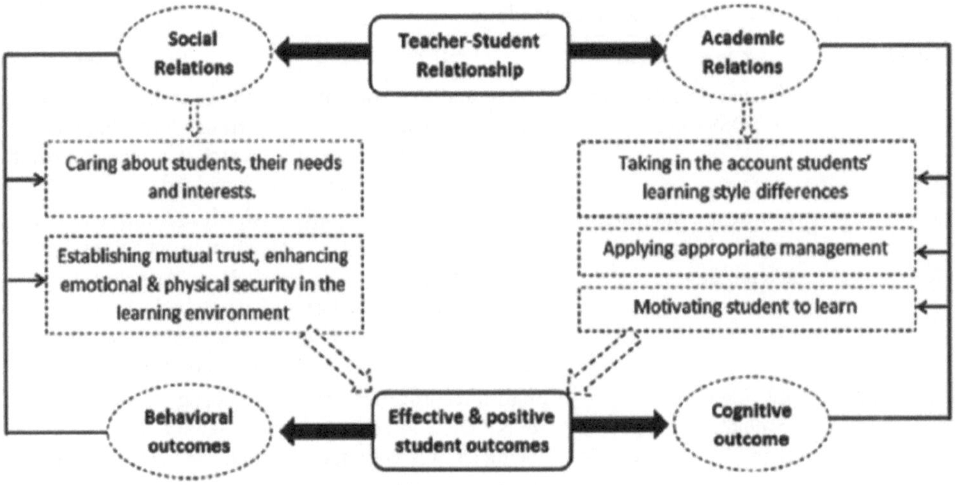

FIGURE 14.1 Teacher–student relationship model.

- Security, ethics, and privacy issues related to the online world need the utmost attention.
- The teacher and learner need to create a health concern and relation for online as well as one-to-one teaching.
- Modern education also witnesses a hindrance due to the socioeconomic divide among Indian people. Artificial intelligence might further deepen the socioeconomic inequalities as a few people (investors) enrich themselves, unless moral factors show the way for us to regulate the economic structure.
- The claim that approximately 4% of AI professionals in India have worked on emerging technologies shows the inadequacies in talent to meet the needs. The collaboration of academia and industry to share ideas and expertise at the higher education level is the need of the hour.
- Faculty lacking technical proficiency creates a severe setback in the present scenario. Innovative schemes like massive open online courses (MOOCs) and SWAYAM should be utilized to the maximum extent to fill this gap.
- The biasing and agendas of the computing systems are in the hands of the programmers (very few) designing algorithms. This poses a serious issue, as implementation is concerned with masses of people. The morality and efficiency issues about technology and artificial intelligence are very important.

14.6 CONCLUSIONS

The essence of teacher–taught relation involves flexibility and rejuvenation of the teaching–learning process so that it could withstand any adverse situation. The hybrid model of education needs to have the right blend of technology and human interaction in the teaching–learning process to have a healthy and flourishing teacher–taught relation. The moral decisions in the implementation of technology involve collaborative efforts of government authorities, engineers, philosophers, economists, and entrepreneurs. Further research is needed to review the role of a teacher, especially in higher education, and embark on a new track of learning with technology to enhance humanism in teacher–taught relations that machines can never replicate.

REFERENCES

1. S. Minocha, A. Schroeder, and C. Schneider, "Role of the educator in social software initiatives in further and higher education: A conceptualisation and research agenda," *British Journal of Educational Technology*, vol. 42, no. 6, pp. 889–903, 2011.
2. S. Deb, "Learning and education in India: Social inequality in the states," *Social Change*, vol. 48, no. 4, pp. 630–633, 2018.
3. S. Ali Shah, "Making the teacher relevant and effective in a technology-led teaching and learning environment," *Procedia: Social and Behavioral Sciences*, vol. 103, pp. 612–620, 2013.
4. K. Subramaniam, "Teachers' organization of participation structures for teaching science with computer technology," *Journal of Science Education and Technology*, vol. 25, no. 4, pp. 527–540, 2016.
5. T. Lickona, Character matters: How to help our children develop good judgment, integrity, and other essential virtues: Touchstone, Simon & Schuster, New York, 2004.
6. All India Council for Technical Education, Government of India. https://www.aicte-india.org/education/IT-and-ICT/
7. G. Mann, and A. Vaishnav, *Committee Report Summary: Draft National Education Policy 2019, PRS Legislative Research*, June 7, 2019. https://www.prsindia.org/sites/default/files/parliament
8. "National Strategy for Artificial Intelligence #AIForAll," *NITI Aayog*, June 2018. https://niti.gov.in/writereaddata/files/documentpublication/NationalStrategy-for-AI-Discussion-Paper.pdf.
9. R. J Seidel, and Ok-Choon Park, "An historical perspective and a model for evaluation of intelligent tutoring systems," *Journal of Educational Computing Research*, vol. 10, no. 2, pp. 103–128, 1994.
10. NPTEL (2017), Retrieved 2017, from http://nptel.ac.in/pdf/nptel/NPTELOnlineCertification-Jan%202016.pdf
11. Spoken Tutorials (2018), Retrieved January 2019, from http://spoken-tutorial.org/
12. SWAYAM (Study Web of Active Learning by Young and Aspiring Minds) programme. (2019), Retrieved 2019, from https://swayam.gov.in
13. T. J. Sabol and R. C. Pianta, "Recent trends in research on teacherchild relationships," *Attachment & Human Development*, vol. 14, no. 3, pp. 213–231, 2012.
14. B. K. Hamre and R. C. Pianta, "Early teacher-child relationships and the trajectory of children's school outcomes through eighth grade," *Child Development*, vol. 72, no. 2, pp. 625–638, 2001.
15. D. L. Roorda, H. M. Koomen, J. L. Spilt, and F. J. Oort, "„e in–uence of aŒective teacher-student relationships on students' school engagement and achievement: a meta-analytic approach," *Review of Educational Research*, vol. 81, no. 4, pp. 493–529, 2011.
16. K. Košir and S. Tement, "Teacher-student relationship and academic achievement: a cross-lagged longitudinal study on three diŒerent age groups," *European Journal of Psychology of Education*, vol. 29, no. 3, pp. 409–428, 2014.
17. A. Forkosh-Baruch and A. Hershkovitz, "Teacher-student relationship and online social networks," in: Tatnall, A. (Ed.), *Encyclopedia of Education and Information Technologies*, Springer, Cham, pp. 1–8, 2019. https://doi.org/10.1007/978-3-319-60013-0_212-1

18. M. E. Brinkworth, J. McIntyre, A. D. Juraschek, and H. Gehlbach, "Teacher-student relationships: the positives and negatives of assessing both perspectives," *Journal of Applied Developmental Psychology*, vol. 55, pp. 24–38, 2018.
19. Z. C. Chan and C. W. Tong, "Power dynamics in the student–teacher relationship in clinical settings," *Nurse Education Today*, vol. 49, pp. 174–179, 2017.
20. J. Bowlby, *Attachment and Loss*, Basic Books, New York, NY, 1969, vol. 1: Attachment.
21. P. P. Liu, M. Savitz-Romer, J. Perella, N. E. Hill, and B. Liang, "Student representations of dyadic and global teacher-student relationships: perceived caring, negativity, aŠnity, and diŒerences across gender and race/ethnicity," *Contemporary Educational Psychology*, vol. 54, pp. 281–296, 2018.
22. L. M. Horowitz, and S. Strack, *Handbook of Interpersonal Psychology: Œeory, Research, Assessment, and Œerapeutic Interventions*, John Wiley & Sons, Hoboken, NJ, 2011.
23. X. J. Sun, H. J. M. Pennings, T. Mainhard, and T. Wubbels, "Teacher interpersonal behavior in the context of positive teacher-student interpersonal relationships in East Asian classrooms: examining the applicability of western findings," *Teaching and Teacher Education*, vol. 86, p. 102898, 2019.
24. L. P. Aultman, M. R. Williams–Johnson, and P. A. Schutz, "Boundary dilemmas in teacher-student relationships: struggling with "the line," *Teaching and Teacher Education*, vol. 25, no. 5, pp. 636–646, 2009.

Index

D

E

J

K

L

M